FOOD AND BEVERAGE SERVICES

FOOD AND BEVERAGE SERVICES

SUNIL SHARMA

AKANSHA PUBLISHING HOUSE
NEW DELHI

AKANSHA PUBLISHING HOUSE
R 37-B, Vani Vihar, Uttam Nagar
New Delhi-110059
Email: ektabooks@yahoo.com

Showroom
4649-B/21, Ansari Road
Daryaganj, New Delhi - 110 002
Phones: 011-23263193, 9811582579

Food and Beverage Services

First Edition 2005
ISBN 81-8370-016-0

PRINTED IN INDIA

Published by M.P. Misra for Akansha Publishing House, New Delhi and
Printed at : Mehra Offset Press, New Delhi.

Contents

Preface

The food and beverage service is an integral component of the hospitality industry. The food and beverage industry is huge and it employs millions of people, more than any other private sector industry. The industry is flourishing day-by day. Now food and beverage services are available in every nook and corner of the world—in large cities, small towns, rural areas, and every place in between.

Today, companies in the food and beverage industry are facing unprecedented challenges. The challenges are driven by issues such as pressure on margins and a marketplace that is crowded with new product introductions in which it is difficult to meaningfully differentiate brands. Issues such as increased automation, never-ending need for improved technology, increased reliability, supply chain management improvements etc., are also challenging the growth of the industry. As these factors and the fierce inter-industry competition continues, it remains a top priority for food and beverage manufacturers to employ adaptable, skilled, and experienced employees.

The combined effect of these challenges is that companies must find innovative ways to create value, mitigate risk, and ensure compliance. One innovative approach to responding to these challenges is improving product traceability across the supply chain. Traceability means the end-to-end tracking of ingredients and chain of custody associated with the manufacture and distribution of food and beverage products. The implementation of a traceability programme will impact a company's strategy, business processes, and technology, and will require a disciplined approach.

This authentic, up-to-date and comprehensive book provides a clear understanding of the fundamental aspects of food and beverage services and the very best strategies for its management. It will prove to be an excellent reference tool for food service professionals and a great training tool for the students or trainees of food and beverage services and hospitality management.

Editor

1

Food and Beverage Operations

Foodservice is one of the largest industries in the world. Foodservices contribute substantially to the mobility of individuals and the flexibility of one's social life. The foodservice industry employs more people than does any other industry. Of every $2.50 spent for food, $1 is spent in foodservices. The industry has enjoyed steady growth since before World War II, with an average increase of about 5 percent per year. Its growth has been the most rapid of that of all retail industries.

The 750,000 or more foodservices that provide this food are widely dispersed and vary from small snack or take-out shops, drive-ins, and coffee shops to moderate-sized restaurants and hotel operations, to huge foodservices such as those serving the military, schools, and large hospital complexes. Airline foodservices alone do millions of dollars in business each year. Because the foodservice industry is labor intensive and because it is divided into many small units that must produce in small amounts, it has never been able to obtain the benefits of mass production.

For more than a decade, beverage marketing corporation has provided superior analysis leading to growth strategy development for beverage industry clients. This in-depth understanding of the beverage industry enables us to bring relevant, market-driven insights and action plans to brand owners, manufacturers, distributors, franchisers and suppliers. Our services are designed to help grow our client's business, with each assignment customized to address the client's unique opportunities and challenges. Our capabilities range from market research and competitive analysis, to developing brand and portfolio growth plans, to strategic and corporate planning.

Control in Food and Beverage Operations

Control means exercising governing power over events and situations such that an outcome can be achieved or prevented. In our personal lives we can normally exercise governing power over our actions, and thus control, to achieve or prevent various outcomes. For example, we leave our homes for work or school each day early enough so that we arrive at a specific time. We drink alcoholic beverages to achieve a desired glow, but we must limit the amount we drink if we do not want to get drunk or sick. Many people give up smoking cigarettes to prevent the possibility of getting lung cancer. Governments, employers, and parents frequently exercise governing power.

It is important to point out that in the food and beverage industry, control really means controlling people rather than things. Food does not disappear by itself, without help. Excess quantities of liquor do not get into drinks unless put there by bartenders. Employees' wage calculations are not based on the wrong numbers of hours unless someone gives the wrong information to the paymaster. Food is not consumed by rodents unless human beings make that food accessible. Customers seldom leave without paying unless staff members make that possible.

In every one of these instances, the problem is the result of human action or lack of it. If a business is to operate profitably and reach its financial goals, people's actions must be managed, or, in some cases, limited. "People" may include more than simply the personnel of the food and beverage operation; they may include customers of the establishment and, in some cases, intruders who seek to steal the resources of the establishment-obviously, without the knowledge and consent of management. Thus, installing locks on both the front and back doors is one of the most basic control measures one can use to prevent intruders from entering and stealing food, beverages, equipment, and cash when the operation is closed.

Locating the cashier near the front door helps prevent customers from leaving without paying their food and beverage charges. Instituting various sanitation procedures may serve effectively to control infestation by various insects and rodents. Obviously, loss of food due to non-human invaders surely constitutes unwarranted additional cost to any restaurant. The time clock for employees is an example of a simple control device. This serves many purposes, one of which is to prepare accurate payroll records so that labour costs will be correct. Another good example is the bartender's use of a measuring device to ensure that each drink will contain the correct amount of a particular alcoholic ingredient.

In the food and beverage business, control is a process used by managers to direct, regulate, and restrain the actions of people so that the established goals of an enterprise may be achieved. Probably the most common goal for all private enterprises is financial success-a particular profit or return on investment Other goals may include operating the best restaurant in the city, having a low labour turnover, or promoting better health by providing nutritional information on the menu. To achieve these goals, management must set up any number of sub-goals compatible with its primary goals. These tend to be more specific and usually more immediate in nature.

For example, to achieve the goal of operating the best restaurant in the city, it might be necessary to have a subgoal of hiring an excellent chef and to purchase the finest ingredients. To have a low labour turnover, it generally is necessary to have a subgoal of establishing a pleasant working environment, good pay, and suitable fringe benefits. A discussion of general business goals is beyond the scope of a text on cost control. Therefore, the focus is on establishing control over costs and sales in food and beverage operations. Attention is centered on the particular methods and procedures used by food service managers to direct, regulate, and restrain the actions of people, both directly and indirectly, to keep costs within acceptable bounds, to account for revenues, and to earn a profit in the process.

Control in Cost

Cost control is defined as the process used by managers to regulate costs and guard against excessive costs. It is an ongoing process and involves every step in the chain of purchasing, receiving, storing, issuing, and preparing food and beverages for sale, as well as training and scheduling the personnel involved. The particular methods used to control costs vary from one establishment to another, depending, in part, on the nature and scope of operations, but the principles behind these methods are constant. The obvious objective is to eliminate excessive costs for food, beverages, and labour-to exert some governing power over costs in all areas to ensure that the enterprise will operate at a profit.

Two of the principal causes of excessive costs are inefficiency and waste. For example, storing food in refrigerators that are not cold enough or bottled beer in a warm, sunny room will cause spoilage and excessive cost. So will the preparation of an inedible beef stew or an unpalatable drink. When the stew is thrown into the garbage can or the drink is poured down the drain, the costs of operation are increased, but sales are not. Because profit is essentially the difference between sales and costs, it is apparent that any increases in cost that do not lead to corresponding increases in sales can have only one effect-a reduction in profits. Clearly, management must take steps to guard against such excessive costs.

Control in Sales

Although cost control is critically important to the profitable operation of any business, cost control alone will not ensure profitability. Additional steps must be taken to ensure that all sales result in appropriate income to the business (sales control). For example, profits will be adversely affected if a steak listed in the menu for $18.95 is sold to a customer for $15.95 or if the drink priced to sell for $4.50 is sold for $2.50, or if food and drink are given away. Therefore, it is important to require that each employee records each sale accurately.

Although it is unfortunate that some employees are not completely honest, it is a fact of life that must be taken into account in food and beverage operations. Therefore, guest checks are usually numbered to ensure accountability. Sometimes duplicate copies of the checks, clearly identified as duplicates so they are not confused with the originals, are used to reconcile kitchen production with recorded sales. When the checks and duplicates are numbered sequentially, missing numbers can be noted and investigated at once. Establishments that use industry-specific computer programs find that control is much easier to institute than at establishments without such programs.

Responsibility for every aspect of any food and beverage enterprise rests with management. Control, therefore, is clearly a responsibility of management. In some food and beverage enterprises-probably the majority—managers take personal charge of directing and supervising the control procedures in every phase of operations. In others, managers delegate some or all of the work to subordinates. The nature, size, and scope of operations help determine the extent to which managers can exercise direct control rather than delegate that control to someone else.

In general, the larger and more complex the operation, the more likely it is that control procedures will be supervised by subordinates. Job titles for these subordinates include food controller, beverage controller, assistant manager, supervisor, and a number of others. In the majority of food and beverage enterprises, one is likely to find control procedures supervised directly by managers, so that title is used most often. In larger operations, one may find food controllers and beverage controllers.

The food and beverage business can be characterized as one that involves raw materials purchased, received, stored, and issued for the purpose of manufacturing products for sale. In respect to these functions, there are many similarities between the food and beverage business and other manufacturing businesses. For example, consider the steps involved in the production of wood furniture. The furniture manufacturer must determine the kinds of wood that are best suited to the types of furniture to be manufactured and must purchase appropriate quantities of lumber at favourable prices. When the wood is delivered, the shipment must be checked to ensure that the material delivered is exactly as ordered. Then it must be moved into appropriate storage facilities, partly to prevent theft and partly to ensure that the characteristics of the wood will not be adversely affected by climate.

The wood must then be issued in appropriate amounts for the production of various kinds of tables and chairs. During the manufacturing process, some effort must be made to maintain balanced quantities of parts so that, for example, one table top is made for each four legs. After the tables have been manufactured and sold to customers, care must be taken to ensure that each sale is properly recorded, that the correct price is charged for each table sold, and that the total dollar value of each sale is collected. The motel coffee shop, the school cafeteria, the hotel dining room, the resort cocktail lounge-all share the same need for purchasing, receiving, storing, and issuing raw materials for the production of finished products for sale.

In food and beverage operations, we deal with fruits, vegetables, meat, poultry, and a large number of other foods-fresh, frozen, and canned-as well as alcoholic beverages in many establishments; but these really serve the same purposes in our business that wood serves in the furniture manufacturing business. Thus, in a very real sense, the food and beverage industry is a manufacturing business. All restaurants take raw material and "manufacture" them into a final product. Some restaurants use raw materials to a greater extent than others. A fine dining establishment, for example, will use relatively few convenience foods. By contrast, fast-food restaurants and institutional operations will generally purchase foods that require less preparation to serve.

In both restaurants and manufacturing operations, raw materials must be carefully selected, always with the desired final products in mind, and appropriate quantities of each must be ordered to meet expected production needs. Orders for materials are typically placed with a vendor at the most favourable prices. As materials are received, they are checked to see that the business is getting what it pays for. Then the materials are stored appropriately until needed.

When the person in charge of production requests materials from storage, they are made available for the production of finished items for sale. At each stage of operation, it is necessary to institute control in order to prevent the kinds of problems mentioned earlier. Control may be accomplished in a variety of ways, and anyone who attempts to manage a food and beverage operation should be aware of the range of techniques available.

Techniques for Control

The control techniques available to a manager include the following:

— Establishing standards
— Establishing procedures
— Training
— Setting examples
— Observing and correcting employee actions
— Requiring records and reports
— Disciplining employees
— Preparing and following budgets

Establishing Quality Standards

Standards are defined as rules or measures established for making comparisons and judgments. In business these standards are set by management and are used for judging the extent to which results meet expectations. Several types of standards are useful in establishing control over food and beverage operations. It is important to develop a working understanding of these several types of standards before proceeding. Quality standards are used to define the degree of excellence of raw materials, finished products, and, by extension, work.

In one sense, establishing quality standards is a grading process. Most food items are graded according to degree of excellence, and management should establish a quality standard for each food item that is to be purchased. Beef, for example, is generally available in a number of different grades for restaurant and institutional use, and it is important to determine which grade will be used for the preparation of a particular menu item. Beverage items also require quality standards. For example, some spirits improve with age, and a 12-year-old scotch whiskey is generally considered to be of a higher quality than one that is only 8 years old.

Management in beverage operations must determine which beverage items are of appropriate quality to ensure customer satisfaction. Quality standards must also be determined for the workforce. In some hotels and fine restaurants, higher degrees of skill are required for the production and service of elaborate menu items. Lower levels of skill would probably be acceptable in the average roadside diner. Quantity standards, defined as measures of weight, count, or volume, are used to make comparisons and judgments. Management must establish a number of quantity standards. Standard portion sizes for food and beverage products and standards for work output are simple examples. The portion size for every food item served must be clearly established. Each shrimp cocktail should contain a predetermined number of shrimp of specified size, a certain measure of sauce, and clearly identified quantities of garnishes.

A portion of soup should be identified as to size of bowl or cup to be used, or the size of the ladle used to portion it, and the quantity of any garnish to be added. Similarly, entrée items must be identified as to the number of ounces or pieces. Surrounding items, such as vegetables, should be portioned with a spoon of a particular size, as with peas, or identified by count, as with asparagus spears. In bar operations, management must establish a standard quantity for each measure of liquor used. In many instances, bars operate with standard drink recipes indicating the specific quantities of ingredients to be used in preparing particular drinks.

Quantity standards are often important in the control of labour costs as well. When planning staff schedules, it is useful to know, for example, the number of tables or seats a server can cover during a given period or the number of sandwiches a pantry worker can make per hour. In addition to quality and quantity standards, it is ultimately necessary to determine and set cost standards for operation. Cost standards are more commonly referred to as standard costs. The term standard cost is defined as the cost of goods or services identified, approved, and accepted by management. Standard costs are used for various purposes. They may be compared with actual costs in order to make judgments about the actual costs, and they may be used as a basis for establishing sales prices. Paradoxically, a standard cost is simultaneously both realistic and an ideal.

Standard costs are useful in measuring the effectiveness of operations in food and beverage establishments. As discussed later, comparing standard costs with actual costs can help determine how effectively food and beverage materials and labour resources are being used in day-to-day operations. Standard costs are particularly useful in cost control because they provide a means for management to compare what is actually happening in an enterprise with what should be happening, given the standards established for operations.

Standard costs must be calculated. In the case of foods and beverages, various kinds of calculations are necessary, the simplest form of which has been previously illustrated in the discussion of the standard cost of 1 ounce of liquor. In the case of labour, determination of standard costs is rather more complex.

Establishing Procedures

In addition to establishing standards for quality, quantity, and cost, food and beverage managers must establish standard procedures. Procedures are the methods employed to prepare products or perform jobs. Standard procedures are those that have been established as the correct methods, routines, and techniques for day-to-day operations. Maintaining effective control over food, beverage, and labour costs requires establishing standard procedures for every phase of operations. Ordering and purchasing procedures must be standardized to ensure that ingredients used to make food and beverage products are purchased at appropriate times in needed quantities, at the most favourably prices, and are of appropriate quality for their intended use. Receiving procedures must be standardized so that all goods received conform in quality, quantity, and cost to those ordered.

Standard storing procedures must be put into effect to guard against both spoilage and theft, either of which will lead to excessive costs. Issuing must be standardized so that food and beverage items will be used in the order in which they are received, thus preventing spoilage and the resulting excessive costs. To further guard against spoilage and theft, the quantities of foods and beverages issued must be linked to carefully determined production needs. Moreover, records of issues can be used to calculate cost per item produced; such costs can then be compared with standard costs to determine the efficiency of production and the effectiveness of operations.

Production procedures must be standardized for a number of reasons. One of the most important of these is customer satisfaction. Any given item should be produced by the same method and with the same ingredients every time it is served. It should also be served in the same quantity each time, partly so that regular customers will be given the same quantity each time they order the item, and partly to maintain cost standards.

Training

Although establishing standards and standard procedures is necessary for control, doing so is really just the first step in a process. None of the standards are of any significance unless employees are aware of them, and employees will not become aware of the standards unless management is willing to undertake staff training. Training is a process by which managers teach employees how work is to be done, given the standards and standard procedures established. For example, if management has established a standard 4-ounce portion size for hamburgers, then all employees responsible for producing portions of hamburgers must be made aware that 4 ounces is the correct portion size.

Moreover, each of these employees must be trained to produce portions of the standard size at his or her workstation, using the correct equipment and supplies provided for doing so. If all employees are not aware of the relevant standards and standard procedures established for their work and are not trained to follow these standards, the standards are completely useless. Any food service manager who trains employees would doubtlessly agree that training is difficult, frustrating, time-consuming, and, in the short run, costly. Perhaps that is why a substantial number of poor managers ignore it and simply put new employees to work without devoting any time to demonstrating and explaining what they are to do.

Sometimes a new employee is merely introduced to a co-worker, who is then expected to train the new person. Occasionally this works reasonably well; more often it does not, frequently because the co-worker is either unwilling or unable to train the new employee. After all, the typical food service employee is not hired for his or her ability to train others. If employees are not suitably trained to follow established standards and standard procedures, the control aspects of the manager's job become difficult, at best; sometimes control becomes impossible.

Setting Examples

Sometimes the process of establishing standards and standard procedures is not quite as formal as the foregoing discussion may suggest. In many instances, standards are established in a very informal way: Employees in an operation follow the examples set by the manager-the manager's behaviour, manner, responses to questions, and even a failure to speak or take action in some situations. In general, the behaviour of individuals in a group tends to be influenced by the actions, statements, and attitudes of their leaders. The attitudes and work habits of a manager are evident as he or she performs various tasks in the course of a workday. The behaviour of the manager will influence the manner in which employees perform their work. If the manager who has occasion to help employees plate food for the dining room serves incorrect portion sizes, employees will be more likely to do the same when the manager is not there.

Similarly, if a manager is inclined to wrap parcels of food to take home for personal use, employees will be more likely to do so. And if the manager observes them doing so and fails to end the practice, the amount of food leaving the premises will usually increase. It must be noted that any manager must be consistent in setting examples, as well as in directing, regulating, and restraining employees and their actions. In far too many cases, managers appear not to have long-range and short-range goals clearly in mind as they go about the business of managing. Consequently, their examples, actions, directions, and responses to employees' questions do not present a clear and consistent view to subordinates. Such inconsistency confuses employees and has the effect of working against the control processes and procedures in effect.

Employees' Actions

If a manager were to see a bartender mixing drinks without measuring the ingredients and did not take the time to remind the individual to measure quantities carefully, then the bartender may reasonably assume that his or her work met the manager's standards. The manager would have missed an excellent opportunity to improve the bartender's work habits and to maintain control. Similarly, if a manager were to observe a receiving clerk failing to verify that quantities of meat delivered conformed to the quantity on the invoice and did not correct the individual, the employee might never know that his or her performance was unacceptable.

One of a manager's important tasks is to observe the actions of all employees continually as they go about their daily jobs, judging those actions in the light of the standards and standard procedures established for their work. If any are failing to follow the standards, it is a manager's responsibility to correct their performance to the extent necessary at the appropriate time.

Requiring Records and Reports

Obviously, no manager can be in all places at all times to observe employees' actions. The owner/manager of a small operation can observe employee actions to a far greater extent than can the manager of a larger operation-one consisting, for example, of several bars and dining rooms, seating hundreds of guests and employing hundreds of personnel. The larger the establishment, the more likely it is that managers' observations must be indirect rather than direct. Their "observations" must be abstracted and inferred from a variety of records and reports. The statement of income summarizes cost and sales information for a particular period. The figures in the statement of income can be used to calculate cost-to-sales ratios of various kinds. These are normally compared with ratios for previous periods, and judgments are made about operations on the basis of these comparisons.

If the comparisons are considered satisfactory to management, this implies, in general, that the performance of employees has been acceptable during the period covered by the statement. If the comparisons are unsatisfactory, the implication is that some or all of the standards and standard procedures have not been followed. A manager cannot normally pinpoint specific problems from such a general report as the statement of income; other more specific and more timely reports and records are usually required. Some of these may be developed daily, others weekly or monthly. At this point, it is important to recognize that managers need timely information to determine whether primary goals and sub-goals are being met. If timely records and reports are not available, opportunities for taking corrective action may be lost.

Discipline

Discipline is defined as action taken to admonish, chastise, or reprimand an employee for work performance or personal behaviour incompatible with established standards. It is a term with negative connotations, suggesting the threats and punishments used on children by some parents. Discipline, therefore, is a difficult topic to address. Food service managers generally prefer to avoid the need for discipline by improving the hiring practices in their organizations. By selecting the right people for the various jobs-those with the experience, skill, and personal characteristics that match the job requirements-the number of individuals requiring some level of discipline can be reduced to a bare minimum. However, every manager must face the fact that at times, an individual staff member must be disciplined. Discipline is used as a control technique in many food and beverage operations and may take any of a number of forms, as discussed in the following paragraphs. It is a valuable technique if used properly and judiciously.

Discipline is not the same as observing and correcting employee actions; it is the next step beyond. Before resorting to discipline, managers are likely to have made reasonable, though unsuccessful, efforts to correct the actions of an employee. Discipline is called for only if corrective action has failed. At this point, managers commonly resort to some words or actions that employees normally regard as punishing, in the broadest sense of that term. One assumes that the standards for work performance and personal behaviour are known and understood by all employees, either because they have been given appropriate training and instruction by management or because they should be aware of the norms for acceptable behaviour in the workplace.

For example, suppose an employee has been trained to follow specific performance standards for serving food in the dining room and does not do so. A supervisor, given his responsibility to observe and correct employees' actions, notices the employee's failure to follow the performance standards and calls his attention to the need to adhere to the standards in which he was trained. If repeated efforts aimed at obtaining improved performance from the employee are entirely unsuccessful, some discipline is required. The employee's poor performance is called to his attention again by his supervisor, probably more directly and forcefully. On this second occasion, the supervisor's approach is likely to be more severe. Most responsible managers hope that the second such occasion will be the last and that the admonition will bring about the necessary improvement.

However, if it does not, then it will be necessary to take progressively more stern, forceful, and, finally, punishing actions, the last of which would be firing. If an employee was worth hiring, he or she should be worth keeping, and all suitable efforts should be made to bring that employee's job performance into line. Therefore, this final step-firing-is normally reserved for use only if it is simply impossible to modify the employee's job performance or personal behaviour. On some rare occasions-in the case of theft, for example-it is used immediately. On another level, a manager responsible for poor financial performance in one unit of a chain organization may have this fact pointed out to him or her by top management in any number of ways.

The possibilities range from quiet discussion to heated admonition, or from demotion or transfer to firing.

Finally, no employee at any level of an organization is completely immune from the possibility of disciplinary action. It is important to recognize the desirability of keeping the number of individuals terminated to a minimum. A high rate of employee turnover because of excessive firings can be very costly, leading to such additional expenses as newspaper advertising for new help, higher rates for unemployment insurance contributions (in some states), lower rates of productivity, higher training costs, and a number of other costs. On one hand, it must be understood that the object of discipline is to change or modify employees' job performance or personal behaviour -to improve performance so that the work is done in conformance with the standards and procedures that management has identified as those most likely to achieve the organization's goals and objectives.

On the other hand, it should be apparent that disciplinary action generally has negative connotations in the minds of employees, most of whom would normally prefer to avoid any such unpleasantness. If certain behaviour patterns lead to positive and pleasant rewards, and others bring the negative and unpleasant consequences associated with disciplinary action, then most employees will generally work toward the former and avoid the latter. The very fact that employees know that continual failure to follow established standards can lead to unpleasantness tends to have the desired effect of making job performance and personal behaviour conform to appropriate standards.

Preparing and Following Budgets

Preparing and following budgets may be the most common technique for controlling business operations. A budget is defined as a financial plan and may be described as a realistic expression of management's goals and objectives expressed in financial terms. Many businesses establish budgets for specific aspects of operation. Thus, there are sales budgets, cash flow budgets, capital equipment budgets, and advertising budgets, among others.

Food and beverage operations also use budgets-many of the same types of budgets that are used in other businesses. Many restaurants prepare capital equipment budgets, for example, because of the ongoing need to replace equipment that wears out and to purchase new types of equipment that becomes available in the market. Capital equipment is typically very costly, and a financial plan must be established before making such purchases. Advertising budgets are established so that appropriate levels of funds can be made available to generate a particular level of sales.

The most important type of budget a food and beverage manager can prepare is an operating budget. Stated in dollar terms, this is a forecast of sales activity and an estimate of costs that will be incurred in the process of generating sales. By extension, the budget indicates the profit that should result after the costs of producing those sales have been met. An operating budget is clearly a financial plan for the period it covers.

For established restaurants, preparation of an operating budget normally begins with historical information-the financial records of the business. In addition, one needs estimates of anticipated changes in costs and sales and in the business environment. Thus, if federal or state laws will affect wage rates paid to employees in the period covered by the budget being prepared, this information must be factored into the planning. Similarly, if the establishment's real estate tax rates have been increased or decreased, that information should be reflected in the budget. More difficult to deal with, but equally important, are the effects of poor weather conditions on future crops, anticipated shortages of beef due to ranchers' reductions in herds, and other such general conditions, including trends in the local and national economy that may affect future costs and sales.

In new establishments that lack financial statements for previous years, information derived from a market analysis and from a manager's previous experience in the industry provide the best alternative sources of data for budget preparation. When no market analysis is available, or when the manager is not widely experienced in the food and beverage business, or when both of these are true, it may be that no effort is made to prepare a budget. When this is the case, management has missed an opportunity to use one of the most useful of all control techniques.

Preparing an Operating Budget

As previously stated, an operating budget is normally prepared using historical information from previous budgets and other financial records. This information, together with anticipated changes in sales and costs, provides the basic data needed to prepare an operating budget for an up coming period. Operating budgets can be prepared for any period of time-a day, a week, a month, a quarter, six months, or a full year. Typically, an operating budget for one full year is prepared first and then broken down into smaller units for shorter time periods. Budgets for these shorter periods may not reflect equal sales and costs in every instance.

This is the case in seasonal restaurants. Thus, for example, if the budget for a restaurant forecasts $1,600,000 in sales for the year, each quarter would not necessarily forecast sales of $400,000. The first and fourth quarters might be busy periods for the restaurant, whereas the second and third quarters might be slow. The manager might forecast $500,000 in sales for the first and fourth quarters and only $300,000 in sales for each of the second and third quarters. In addition, the manager might forecast substantial profits for the busy quarters and very little for the slow quarters.

Handling of Food

A basic plan of operation will be useful in determining necessary equipment and standard procedures so that you offer safe foods for sale. Safe foods involve controlling temperature, contamination sources, and time at improper temperatures allowing microbial growth. A plan of operation sets out what you want to do and how to accomplish your goals. Having it written out, sets guidelines for employees and is a reference for you in evaluating success and future improvements. Your plan will follow the path the food takes to reach sale. It will include when and where you get the product, how and where it will be displayed, any special handling or protection needed, hours of operation, and what will be done with unsold product.

It is most important to note the critical points where control is necessary to keep food safe. Although each food is different, the common critical control points are:

— Food temperature

— The time potentially hazardous food is held without refrigeration or heat

— Potential paths of contamination/adulteration

The complexity of the operation is based on the foods' potential to cause food-borne illnesses. Fresh whole tomatoes are not potentially hazardous and are safe to sell without packaging, refrigeration, or even an overhead cover. Frozen meats, are potentially hazardous and must be carefully processed and packaged, labeled, held under refrigeration, and protected from the weather. The potential hazard level of each product dictates the operation as well as the number and kind of critical controls. Other considerations might include:

- — Is the operation seasonal or year round-What about weather protection?
- — Other time demands-if your time is limited, the operation should be simple.
- — Location of processing and sale-Are surfaces paved? Are animal pens nearby?
- — Employees-if needed, are they trained and supervised?
- — Are processed foods sold, or does any processing occur at the market place?

Simple seasonal operations like selling freshly harvested tomatoes or ears of corn function very well without power, running water, etc. If you decide to become a licensed retail market, utilities will be required. Your plan of operation should alert you to utility needs. Be sure to confirm availability of utilities before committing time and money to establishing your facility. In general, a safe water supply and an acceptable wastewater treatment method are necessary when food processing involves water, or when sinks and toilets are in use. In some cases, the water supply may be required to meet public water supply requirements. A source of power is required for refrigeration, lighting, and machinery operation. Be sure to consider waste management so that insects and rodents are controlled.

Storage and Display

All foods are better protected if temperatures are controlled. Even canned goods and dry products such as herbs or grain meals and f lours retain quality better at moderate temperatures than at high temperatures. Raw fruits and vegetables that have not been cut up or otherwise altered can be displayed without refrigeration if air temperature is above freezing. Fresh produce quality is generally prolonged by cool conditions. Processed meat like dressed poultry, must be kept cold at refrigeration temperatures

(40°F or less) until delivered to the customer. Frozen meats must be held at or below 0°F. In general, foods that have been processed and contain moisture are considered potentially hazardous and are more at risk from growth of bacteria or fungi.

Pathogenic bacteria need several conditions in order to grow in foods: adequate pH, oxygen (or lack thereof) moisture, temperature, and time to reproduce. Pickling foods changes the pH to an unacceptable level for many bacteria. Replacing oxygen with a vacuum or another gas stops the growth of most types of bacteria; however, the bacteria that cause botulism grow in the absence of oxygen. One way we can stop bacterial growth is by removing moisture, such as when drying herbs. Another way to control growth is with temperature. Most bacteria cannot grow if the food is either quite hot (at least 140°F), or (below 40°F). The last growth factor-time-relates to foods kept at temperatures between 40°F and 140°F. We can still control growth by limiting the time potentially hazardous food is held in this temperature range to no more than two hours.

Another consideration is the number of microbes present. Most pathogens cause illness only after a minimum number of organisms have been produced. That number can range from one Cryptosporidium, to many thousands of Streptococci. If the food has been contaminated with only a few organisms, it takes longer to reach the numbers necessary to cause illness. When many organisms are present, it takes less time to reach this dangerous number. Anything that decreases the amount of bacteria therefore improves the food's safety factor. Preventing microbial contamination makes food safer. Consider a cantaloupe melon; as long as the rind is undamaged, the interior is uncontaminated and the melon is not potentially hazardous. When the melon is cut open, it becomes potentially hazardous because the interior is exposed to bacterial contamination. Therefore, once cut open, the melon temperature must be kept below 40°F to control bacterial growth.

Some processing techniques may physically alter the food or remove natural barriers to contamination, making them potentially hazardous. Cooking alters vegetables by weakening cell walls, allowing microbes to grow better. If the variety and number of microbes change through the processing -as happens when cooking kills all bacteria-the natural competition that normally limits growth is altered. Should disease causing bacteria then contaminate the food, multiplication is unhindered by barriers or competition, and the food becomes dangerous more quickly. This is why keeping cooked foods safe involves controlling both time and temperature.

You may think you can judge temperature by feel or by seeing steam, but the only sure indicator is a thermometer inserted into the food. As mentioned before, foods that have been heat treated or physically altered, are potentially hazardous and must be kept hot (at least 140°F) or cold (below 40°F). It is a good idea to document food temperatures, with date and time on a log sheet. Whenever more than one person is involved in processing temperature sensitive foods, there is no other way to know what the temperature history is.

During display it is necessary to have some means to maintain the appropriate temperature. Although insulated containers slow heat movement, remember the two-hour time limit for food out of temperature. Any time potentially hazardous foods will be out of temperature for more than two hours, mechanical devices are best to safely hold the food at the proper temperature. Be sure there is a stem thermometer to check food temperatures and a thermometer to monitor the unit's temperature.

Sometimes crushed ice is used to maintain cold temperatures. In this case, the crushed ice should be in an insulated, lidded cooler, with a drain in the bottom for discharge of melt water into a bucket. Temperature sensitive foods should be surrounded by ice and not simply set on top. If the food is ready to eat and will not be cooked, it should be inside a covered container bedded in ice.

An attractive display enhances sales and should keep food safe. Temperature control has been discussed, but remember that the table, bin, or rack that contacts the food may be laden with microbes. Therefore, all such food contact surfaces must be kept clean and should be easily cleanable. Contamination risks can be decreased by dedicating the bin or container to the type of crop. For example, the same container should not be used for root crops and fruits. Kansas regulations require that foods be at least 18 inches off the ground for sidewalk (outdoor) display. Wrapped products sold intact, are protected from exposure to dust and insects. Baked goods should always be prewrapped for sale and never portioned at the sale.

Live animals increase the risk of contamination of nearby food displays. Not only does the animal manure attract f lies, but blowing dust from their pens or cages can carry microbes to the food. If the community open-air market includes vendors selling live animals, those vendors should be grouped and separated from food booths. Animals should be down wind from food displays, and as far away as possible. Privately owned markets on the farm site should be located up wind and well away from animal pens.

Cleanliness

Rough wooden tables and bins are difficult to clean, making them unsuitable for direct contact with some food products. Wood surfaces should be smooth and sealed with enamel or paint for ease of cleaning. Washable or disposable covers and liners are additional protection for the food, and make clean-up simpler. They present a more attractive display too. You have control over the use and storage of your tables and display bins. It is best to dedicate these items for like products such as tubers, or fruits grown off the ground, or processed foods. This means that the bin used for potatoes is not later used for displaying apples. Display items should be stored where they cannot be contaminated by manure or chemicals.

2

Food Service Operations

Foodservice operation is the overall plan for how it will meet the needs and expectations of its intended market. A foodservice operation's concept is expressed in many ways, including its menu, decor, form of service, pricing, and location. Concept development means developing a plan for the success of the operation in its market in advance of actually designing -let alone building-the facility. It is not unusual for a person to consider a new restaurant or, in fact, to open a new restaurant without knowing what type of food facility will have the best chance of succeeding.

The potential entrepreneur may have some investment money, a location or a theme in mind, and a great amount of enthusiasm for the food business, but has not really thought through the total concept of the operation. Unfortunately, enthusiasm and great food products are only half of the success equation. The other half of the equation is the market. Concept development precedes the actual design of a foodservice facility because the foodservice design team must know what the menu, demand, hours of operation, and mode of service will be.

Single-Unit Restaurant

The client who most frequently comes to the food facilities design consultant for help with concept development is the individual restaurant owner. The restaurant owner typically organizes a corporation comprised of a small number of local businesspeople and then begins to develop a concept that will eventually become a freestanding restaurant. The success or failure of the venture often depends on how well the concept was planned and how well the plan was followed.

Numerous concepts are possible for single-unit restaurants. Commonly found concepts often are described in terms of the following general categories:

— *Fine-dining restaurants* : Fine-dining restaurants are distinguished by fine cuisine prepared by celebrity chefs, attentive service, stylish decor, and high prices.

— *Theme restaurants* : Theme restaurants offer a dining experience that evokes special times, places, or events, such as English pubs, restaurants owned by sports celebrities, and re-creations of diners from the 1950s.

— *Casual dinner houses* : Casual dinner houses emphasize a comfortable and contemporary decor, as well as high value. Well-known casual dinner houses are not single-unit restaurants, but chains such as Bennigan's, T.G.I. Friday's, and Max & Erma's.

— *Ethnic restaurants* : Ethnic restaurants are closely tied to the cultures or foodways from which they originated. They include Mexican, Italian, French, German, Thai, and Indian restaurants, to name but a few.. Family restaurants. Family restaurants specialize in relatively inexpensive fare and are kid-friendly.

— *Quick-service restaurants* : Quick-service restaurants specialize in convenience and fast service and include fast-food operations as well as delis, bagel shops, and sandwich shops.

Each of these categories of foodservice concept involves differences in menu, decor, mode of service, and price. However, not all of these factors are equally important within a given concept. Price is a critical factor in the success of quick-service, family, and casual dinner restaurants, where customers are value-conscious. However, price may not be as important in finedining restaurants, where customers expect to pay top dollar. Similarly, location is crucial for quick-service restaurants because their clientele depend upon convenient access. But for some fine-dining and theme restaurants, location is not critical. Concept development for a single-unit restaurant is thus a complex process.

Chain Restaurant

When Dave Thomas, the late chairman of the board of Wendy's, traveled around the country with Colonel Sanders in the mid-1950s trying to promote a chicken franchise, he learned many of the dos and don'ts of food franchise marketing. Thomas certainly picked up good ideas about concept development for chain restaurants and franchises, as the success of Kentucky Fried Chicken (known as "KFC") and then Wendy's demonstrates. The basic objectives he developed, which led to the formation of Wendy's, were the following:

— Produce a "Cadillac" hamburger with a large number of available condiments.

— Limit the menu to the smallest number of items possible, as most restaurants can prepare only a few food items extremely well.

— Create an image different from major competitors. In the case of Wendy's, distinctive features included an old-fashioned, nostalgic theme, carpet on the floor, marketing directed at adults, and a larger hamburger than the competition's.

Concept development for Wendy's was more comprehensive than that for a single-unit restaurant. The franchise strategy was carefully thought out to create a balance between company-owned stores and franchised stores. In 1970, only two stores were open, both of them owned and operated by the company. By 1975, 83 company-owned stores and 169 franchised stores were in operation. A ratio of 30 to 40 percent company-owned stores to 60 to 70 percent franchised stores permitted a balance of control and greater financial return. The strategy entailed rapid expansion of the franchise and heavy promotion of the Wendy's name through national advertising. Wendy's now has over 5,400 stores worldwide.

Multiunit casual dinner houses and theme restaurants, such as Chili's, Bennigan's, and the Olive Garden, follow a similar pattern in concept development. Their emphasis is on identifying the key characteristics of their target markets and then locating restaurants where there is a high concentration of individuals who have those characteristics. Key characteristics may include income, age, education, and home ownership. What gives multiunit restaurants a competitive advantage over singleunit restaurants is the opportunity to learn from experience with multiple examples of the same concept. When a chain restaurant firm has five hundred virtually identical restaurants, it can analyze the factors that differentiate its high-performing restaurants from its low-performing restaurants, and make changes as necessary before opening additional units.

Multitheme Restaurant

A particular form of multiunit restaurant for which concept development is critical to success is the restaurant organization that opens and operates restaurants whose concepts are not identical but different. The Levi Organization and Lettuce Entertain You are two examples of successful restaurant chains that have developed multitheme restaurant concepts. These two companies each use several different themes, and each restaurant is promoted with its theme rather than by using the corporate name. The development of these restaurant concepts through excellent marketing, well-planned menus, and good design comes about through the efforts of a very sophisticated management team.

Noncommercial Foodservice

Noncommercial (Institutional) foodservice is usually conceived as a service to an organization, and most often has a not-for-profit philosophy. Most institutional food operations are expected to break even, and all are expected to budget and operate within well-defined ranges of costs so that they do not become a financial burden on the organization they serve. In some cases, the institutional food operation is expected to make a profit and to pay for all of its direct and indirect operational costs. The development of an operational concept for the institution is often ignored, and this is usually a serious mistake. The institution must accurately interpret its market and must "sell" its products, even when the food is indirectly paid for by the customer.

For instance, in hospital foodservice, an unattractive meal presentation will cause dissatisfaction and complaining on the part of the patient and possibly adverse health effects as well if he or she does not eat a meal and thus does not get sufficient nourishment. In a college or university dining hall, a comprehensive concept of service and decor can greatly influence financial success. Attractive food court service or a scramble design, for example, can increase the popularity of a college foodservice operation and generate additional profit. A dining facility operated by a corporation for its employees should also have a well-planned concept and decor. The ability of corporate foodservice operations to attract employees may influence the degree of subsidy that a company is willing to contribute to the operation.

Five M's for Concept Development

The successful foodservice operation combines the following elements of concept development: market, menu, money, management, and method of execution (Figure 1).

Market

The importance of conducting market studies before proceeding with the construction of a food facility cannot be too heavily stressed. A classic mistake made by both large corporations and individual restaurant operators is to conduct the market analysis and then fail to act on the basis of the information obtained. There are several cases in which extensive marketing feasibility studies were conducted by outside marketing firms, but the owners and managers made their decisions on gut feelings rather than from the hard data derived from the study. Even owners (or potential owners) of food operations who have no marketing background can conduct their own market research, with a small amount of guidance and a large amount of energy and common sense.

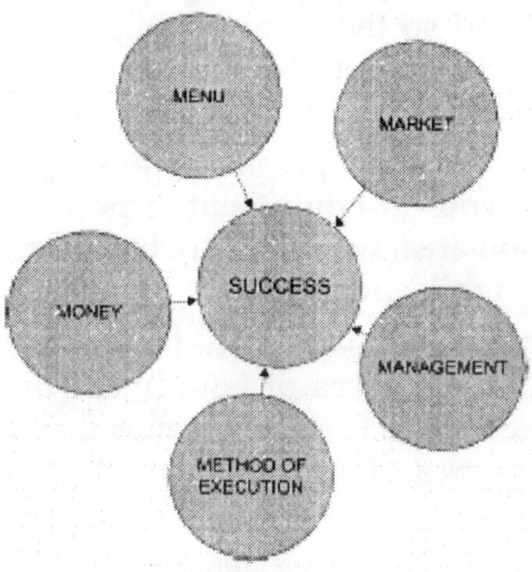

Figure 1. Five M's of concept development.

Menu

The importance of the menu to the design of the food facility cannot be overemphasized. The subject of menu writing is too broad to be addressed adequately in a book on foodservice facilities design. The owner or manager is encouraged to seek additional sources of information as a part of the process of developing a menu for a new or renovated food operation. The menu has a tremendous influence on the design and success of a food operation. From a design and layout perspective, these are just some of the factors determined by the menu:

— *Amount of space required* : A complex menu requires more space to prepare than a limited or simple menu does, because separate workstations and additional equipment are necessary.

— *Service area size and design*: The greater the number of menu items, the more area required for service. For example, in a cafeteria each beverage requires a dispenser and each entree a point of service.

— *Dishwashing area size and dish machine capacity*: Complex menus often require multiple plates, dishes, and utensils, and so the dishwashing area and machine capacity will need to be greater than in the case of simple menus.

— *Types of cooking equipment*: Complex menus require multiple types of equipment, especially in the final preparation area, where it may be necessary to steam, fry, bake, broil, and sauté.

— *Equipment capacity*: Limited menus may require relatively few pieces of equipment but need large capacities of each. Complex menus may require many different types of equipment with relatively small capacities.

— *Size of dry and refrigerated storage areas*: Complex menus may require larger storage areas to maintain the par stocks necessary to meet demand.

— *Number of employees*: Simple menus require fewer employees than do complex menus.
— *Amount of investment required*: When large or complex menus require more equipment, space, and employees, costs rise.

Money

Successful capitalization of a food facility includes funds for:

— Planning costs
— Building construction or renovation
— Equipment (fixed)
— China, glassware, utensils
— Furniture and fixtures
— Decor
— Operating costs

These funds must be identified and committed before serious planning can begin. Yet, in concept development, the commitments may not be made in the early planning stage because the costs are not yet known. Therefore, planning for capital funds is a two-step process: First the financial needs are estimated and sources of financial support are contacted to determine the possibility of obtaining investment funds; then, after concept development has taken place, preliminary designs and construction estimates have been made, and market research is completed, financial commitments are made by lenders and investors.

Management

The quality of the management of the foodservice operation will be the most important element in achieving success. The successful restaurant often is owned and operated by one individual whose personality becomes a part of the guests' dining experience. On the other hand the management of the food and beverage department of a hotel may be under the control of more than one person and usually is part of a more complex organizational team. In this case, the policies and procedures of the food facility should be described in an operations manual to ensure consistent implementation of management policy.

From the point of view of the investor or the institutional administration, the management of a food facility must follow traditional management principles of good communication, strong controls, and sound personnel relations regardless of the number of people operating the facility. The operational philosophy and specific management guidelines to be used in foodservice operations must be carefully considered by the investors in a foodservice facility. Failure to develop management guidelines will very likely lead to the financial failure of the operation.

Method of Execution

The last step in concept development involves operational matters. Although the opening date might seem to be in the distant future to the person planning a food facility, decisions about operating methods must be made during the concept development phase on matters such as production methods, control systems, and personnel.

Production methods

Will convenience foods or traditional "from scratch" cookery be used? This decision will have a great influence on the size of refrigerated and dry storage areas and on the size of the kitchen. Production methods will also determine the number of employees in the kitchen and the skill level of these employees.

Control systems

Food and beverage controls involve many different parts of the facility, and planning for these controls before the project is under construction is strongly recommended. The following areas of control should be carefully considered:

— Cash control
— Sales analysis
— Guest check control
— Food production forecasting
— Storeroom and refrigeration control
— Back door security
— Labor control
— Purchasing and receiving control
— Quality control
— Portion control

Personnel

The development of financial feasibility studies cannot begin until the amount of labor required is known. The employee schedules, hours of operation, staffing patterns, staff benefits, skill levels, and level of supervision of employees must all be determined before serious development of the food facility begins. As part of its concept development, the fast-food industry based its low labor costs on the use of hourly unskilled labor, scheduled to work short periods of time. When the food operation is busy, part-time employees are scheduled to work. The traditional eight-hour day is seldom used in the fast-food industry, except for supervisors and managers. The use of part-time employees in fast-food restaurants has also significantly reduced the cost of benefits. The use of part-time employees was an important part of concept development in the fast-food industry.

Feasibility

Many terms are commonly used in the hospitality industry to describe the process of determining whether or not a food facility is likely to return a profit to its owners. The following is a partial list of these terms:

- Market or marketability study
- Market segmentation analysis
- Market and operations analysis
- Appraisal report
- Economic study
- Time-share feasibility study
- Feasibility study, report, or analysis
- Financial feasibility study
- ROI (return on investment) analysis
- Sales/performance study

Although each term has a slightly different meaning or involves a slightly different approach, they all share the goal of determining the potential of a facility to generate sales and a profit. In the case of the financial feasibility or ROI (return on investment) analysis, the emphasis is on financial matters such as capital needs, operating funds, cash flow, and return on investment. However, even financial feasibility reports have as their primary focus the determination of whether or not a facility under good management can give investors or owners a return on their investment.

For purposes of explaining the feasibility studies and of guiding the owner, manager, or student into a commonsense approach to these studies, they are classified here into two general categories: those that deal with market feasibility and those that attempt to determine financial feasibility. The two forms of feasibility study can be understood in relation to the financial statements for the operation. The market feasibility study focuses on the income statement and is conducted to determine whether revenues are sufficient to generate a profit. The financial feasibility study focuses on the balance sheet and is conducted to determine whether retained earnings (derived from net income) will be sufficient to satisfy the owners' expectations for a return on their investment.

Market Feasibility

The market feasibility study attempts to project the sales level for the operation before the investors have committed their funds to purchase property, construct a building, and hire a team of employees. The sales revenue for a foodservice facility is a function of two factors over an appropriate period of time: the number of customers and the price they paid.

$$\text{Sales} = \text{Price} \times \text{Quantity}$$

Market feasibility studies thus have to formulate sales estimates from two separate projections: How many customers will there be? How much will each customer spend? These two estimates are interrelated through the simple economic principle that demand (quantity sold) is more or less a function of price Market feasibility studies can be conducted either to test the feasibility of an established foodservice concept or to develop new foodservice concepts that would be appropriate for a given market.

Chain restaurant organizations interested in expansion usually want to determine whether, or where, in a given market area their concept would enjoy the greatest opportunity for success. Independent restaurateurs, as well as chain restaurant organizations, who are interested in developing a new restaurant use the market feasibility study to develop and refine the concept. Because the market feasibility process is complex and time-consuming, it is often conducted by specialized consultants.

Independent restaurateurs may find that the cost of engaging a market feasibility consultant is prohibitive. However, it is possible-and certainly advisable-for entrepreneurs to conduct an abbreviated market feasibility study. Resources and guidelines for conducting a restaurant feasibility study are available from the National Restaurant Association.

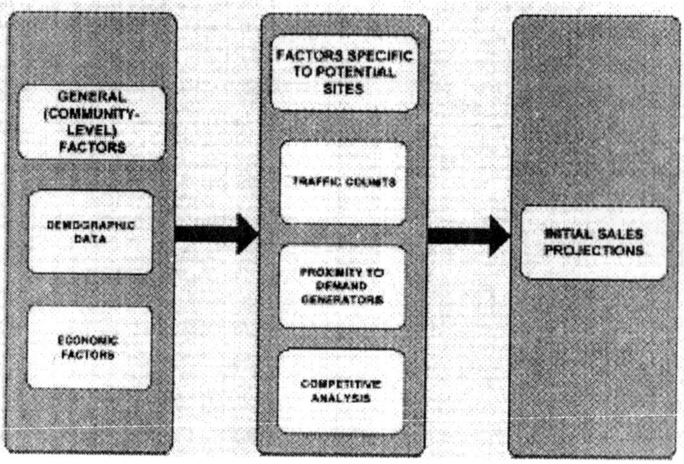

Figure 2. The market feasibility process.

Market feasibility studies generally follow the process presented-in simplified and shortened form-in Figure 2. Demographic data and economic factors for the community are analyzed to determine the potential market for the restaurant. Traffic counts and proximity to demand generators are used to assess the viability of possible sites for the restaurant. A competitive analysis is performed to determine whether the market can sustain another restaurant at each potential site. Based on these analyses, sales projections are formulated. Each of these steps in the market feasibility study described below.

Demographic data

Demographic data include information about the population near the proposed location of the restaurant, including age, income, gender, occupation, number of children at home, home ownership, the use of automobiles, and similar information. These data about individuals are relevant because target market segments for restaurants are often described in terms of the same factors. Demographics thus are known to predict restaurant behavior. For example, the developers of a casual dinner house may believe that their concept will attract young, single professionals between the ages of twentysix and thirty-four who have incomes in the $35,000-$50,000 range and who pay monthly rents between $1,000 and $1,600.

Obviously, the dinner house should be located where there is a high concentration of individuals who fit the profile of the target market. Gathering demographic data enables the developers to detect the presence of the proposed restaurant's target market segment. The most comprehensive demographic data are gathered in the decennial U.S. Census. Until recently, using census data involved either extensive library research or access to a mainframe computer. Now, however, detailed census data are available to the public on CD-ROM at a reasonable cost.

Also, in the interest of promoting economic growth, many local municipalities have developed extensive databases combining demographic data with information about businesses, organized on detailed maps of the area. This kind of database, called a GIS (geographic information system), gives restaurant developers immediate access to information about the market and the competition. GIS software and databases also are increasingly available for desktop computers and so bring the resources of high-priced market consultants to the fingertips of entrepreneurs.

Economic climate

The prevailing economic climate in a community has a significant impact on the market feasibility of a foodservice operation. The health of the restaurant industry in a local area is closely related to the disposable income of its residents. Generally, when disposable income rises in boom times, restaurants prosper. In bust times, disposable income falls and restaurants suffer. However, the effects of economic prosperity and depression affect various restaurant concepts differently. Concepts that offer a high degree of perceived value for the price will suffer less than restaurants with high prices. Also, restaurants that draw customers from a particular industry or from a few large, nearby firms may be protected from broader economic fluctuations if the industry or the large firms prosper during down times.

Economic projections for a local area often are available from the chamber of commerce or the local economic development council. New housing developments are an important indicator of economic growth, especially when they are located on the fringes of existing communities. New restaurants often follow new housing developments. County planning commissions usually are quite willing to share information about planned housing developments, new roads, and zoning issues.

Traffic counts

Restaurants are usually developed along major thoroughfares. Determining market feasibility requires an estimate of the number of cars that pass by each potential location, which is used in determining which site is optimal. Traffic counts can usually be obtained free of charge from the local chamber of commerce, highway department, mayor's office, tourism agency, or other municipal offices. Some fast-food chains have developed very exact specifications for locating a good site for the foodservice facilities. For instance, one of the guidelines for placement of a Kentucky Fried Chicken (KFC) outlet is that it be located on the "going-home side of the street," for obvious reasons. Other chains have minimum community size, minimum traffic count, and minimum average income standards that must be met before consideration will be given to building or franchising on a particular site.

Demand generators

Demand generators are destinations that draw extensive traffic, such as shopping malls, recreation and sports facilities, and public facilities such as museums, zoos, and parks. Restaurants located near demand generators benefit from the traffic they produce-as, for example, when shoppers have lunch after visiting the local mall. For some foodservice concepts, proximity to other restaurants enhances rather than hurts market feasibility. Fast-food concepts, for example, often locate in clusters along interstate highway interchanges. Similarly, cities and suburbs often have a "restaurant row" in commercial areas.

Competitive analysis

A competitive analysis is crucial to the market feasibility study because the local market potential for a given concept must be shared among all of the competitors. If demographic research identifies an area that has a sales potential of $6 million to $8 million for casual dinner houses, it is important to know how many already are in operation. Moreover, when the restaurateur identifies an area that appears to be ripe for development, he or she should assume that the competition will soon arrive. Most chain restaurant organizations have access to GIS market data, and many of them use those data extensively to identify prime development opportunities.

An independent restaurant entrepreneur who is conducting a market feasibility analysis should identify all the similar concepts within the geographic area, compare menus and prices, and count seats. More importantly, he or she should count customers at various times during the week at the restaurants that would be the closest competition. If patrons are lined up outside the doors to get a seat on Monday or Tuesday night, that is good news. However, if there is no waiting at the competition on Friday or Saturday night, the independent restaurateur should look elsewhere. It is also important to take seasonal factors into account. Counting cars and customers at the competition during the summer months will bias estimates upward, just as counts taken during the winter when two feet (610 millimeters) of snow are on the ground may bias estimates downward.

Sales projections

Sales projections are informed by the information gathered about demographics, economic factors, traffic patterns, demand generators, and competitive analysis. Recall that sales projections are composed of estimates about quantity (number of customers) and price. Demographic data indicate the relative size of the local population and thus are useful in projecting the number of customers (quantity). Income information may also be found in demographic data and may thus assist in estimating the price range that people are willing to accept. Economic factors are especially important in projecting price. Entrepreneurs often make the mistake of basing their sales estimates on data gathered during boom times and thus find themselves suffering during bust times. Traffic pattern data are crucial in estimating customer counts, as is the presence of demand generators.

Competitive analysis helps the entrepreneur determine his or her share of the customer market and predict price points against the competition. The weakness in many market feasibility studies can be found within the sales projection section. This part of the feasibility report is often the first place that a banker will look to determine the accuracy of the financial forecast. Bankers always appreciate conservative sales projections that reflect the restaurateur's planning for sales fluctuations during difficult periods. Restaurant entrepreneurs also often overestimate the dining patterns and price sensitivities of the local market. Sometimes this is the result of wishful thinking, but other times it is the result of insufficient planning and preparation.

A valuable way for the entrepreneur to check his or her assumptions about the market is to develop a questionnaire and distribute it to those who reside or work in the area. A questionnaire is helpful in determining such things as:

— Detailed demographic data about age, income, and family size

— Eating-out patterns (frequency, meal preferences)

— Price sensitivity and the average price paid for lunch and dinner at competing establishments

— Favorite eating places for various occasions

For a hotel, the sales pattern would be influenced by the house count. The calculation, therefore, should be based on the projected occupancy. Data on hotel occupancy levels are available from leading hospitality industry accounting firms. Price Waterhouse Coopers, Smith Travel Research, and Pannell, Kerr, Foster, for example, regularly publish hotel operations data from all parts of the country and different segments of the hotel field (resorts, transient hotels, and motels). By utilizing these resources, one should be able to make accurate sales projections that will be credible to the lending institution or investor.

Projections for hospitals and nursing homes are based on patient room occupancy data, but in the health care field a feasibility study usually does not include any data on sales. Foodservice in hospitals and other health care institutions is a service and support arm of the facility, with the primary management and financial considerations focused on the quality of the service and cost containment of the operation. Once relevant market data have been gathered, a complete sales forecast should be prepared.

Financial Feasibility Study

The lending institution, investor, owner, and manager will all want to know the financial projections for the planned new or renovated food facility. Each of these persons will, of course, have a different set of reasons for seeking the projections, and each will want data from the projections presented in a different manner. For instance, the banker will be looking in part for the ratio of invested capital to borrowed capital. The banker may also want to know the amount of operating cash and the cash flow from sales that will be involved in the operation. The manager needs to know what his or her budget is and what the expectation of the owner is concerning profit and loss.

The manager and the banker probably will not be using the same financial reports and projections, but they certainly will be getting their information from the same original source, which will probably be the financial feasibility study. Usually in the financial feasibility study two basic documents -the projected balance sheet and the pro forma profit-and-loss statement-are prepared, along with other supporting reports and schedules. A good outline of the kind of information that must be projected can be made by examining the line items on each of these documents.

Balance sheet

A simple comparative balance sheet illustrates the kind of

data that must be determined. This balance sheet is a simplification of the projected comparative balance sheet that should be developed under the guidance of the firm's accountant. Note that the balance sheet shows a projected comparison between the assets, liabilities, and capital for a twelve-month period. The balance sheet illustrates a number of projections that must be made by the person or firm preparing the financial feasibility study. On the asset side of the balance sheet:

— The amount of cash needed as operating funds
— The amount of cash tied up in food inventory
— The investment in land and building
— The cash needed for a down payment on land and building
— The cost of furniture, fixtures, equipment, and utensils
— The cost of parking lots, driveways, lighting, and other improvements to the property

On the liabilities and equity side of the balance sheet:

— Necessary short-term funds that must be borrowed
— The amount of payables
— The amount of long-term mortgages on building and land
— The amount of funds that will represent the owner's equity (capitalization)

Decisions regarding the type of business organization (proprietorship, partnership, corporation, and so forth) and control of ownership (closely held, limited number of investors, the sale of common stock, and so forth) are made prior to preparing the projected balance sheet.

Pro forma profit-and-loss statement (P&L)

The P&L statement should be prepared by the owner, manager, or other persons who will be involved in the management of the food facility. This statement projects the income and expense for a particular period of time. For the financial feasibility study, a three-to-fiveyear projection would be considered sufficient. The format for the P&L should follow the Uniform System of Accounts developed for hotels and restaurants for income and expense categories that fit the needs of the food operation. Several supporting schedules are prepared to supplement the P&L statement. Because food and labor costs combined typically run between 60 and 80 percent of sales revenue, schedules providing detailed information on how the cost of food sales and the cost of personnel and related expenses were calculated are often provided.

Cost of Sales: A schedule detailing how the amount shown on the cost-ofsales line of the P&L statement was calculated is often prepared. Because the cost of sales for the menu equals the selling price minus the cost of the ingredients used in preparation, two closely related steps are involved in preparing the schedule: determining the portion cost for each menu item and estimating the price for each menu item. The portion cost for each menu item is determined by examining the recipe and costs for each ingredient. The form provides the following information:

— A list of all food ingredients

— The portion size of each ingredient

— The cost of each ingredient

— The total portion cost

If the menu price includes the meat, vegetable, dessert, and beverage, then the above example would necessarily include a complete list of all foods that are a part of this meal. A small amount ($.05) might be added to cover the cost of seasonings and other condiments. The two traditional methods for determining the selling price for menu items are (a) to divide the portion cost by a set percentage (the desired food cost percentage for the operation), and (b) to add a set dollar amount to the cost. Using the first pricing approach, a manager who seeks to have a 32 percent food cost percentage would divide the food cost of a menu item by the desired percentage. For example, a food cost for a ham sandwich of $.72 would be divided by.32, and the selling price would be $2.25. Using the second approach, the manager would add a fixed amount to the cost of each menu item. Adding $2 to the cost of the ham sandwich would result in a selling price of $2.72.

A third approach to setting prices involves looking not at each item but at the menu and menu mix (the quantity sold of each menu item) as a whole. The goal of this approach is to set prices so that the contribution margin (the amount of revenue remaining when the cost of sales has been subtracted from revenue) is sufficient to pay all expected fixed expenses and satisfy the owner's expectations for a return on investment.

The advantage of the spreadsheet approach over traditional strategies to pricing, such as simply marking up the cost by a given percentage or dollar amount, is that prices can be set to generate the maximum contribution margin. In foodservice, profits come not from minimizing the food cost percentage but from maximizing the contribution margin. However, for the purposes of preparing the pro forma profit-andloss statement, the overall food cost percentage derived from the spreadsheet can be used to estimate the cost of sales.

Finally, it is essential to check the competition in setting prices for each menu item, because an operation that is not competitively priced will suffer. Once the menu pricing and preferred menu mix have been established, the calculation of the cost of food sales is simple: The projected sales revenue is multiplied by the overall food cost factor, and the result is shown on the pro forma P&L statement.

Cost of Personnel and Related Expenses: The method of projecting the total labor costs for the pro forma P&L

statement is to decide on the number of persons needed in each job category and then prepare an employee sched-ule. Salary and wage levels should be estimated, and the annual cost of labor computed. A small amount, perhaps 10 to 12 percent, should be added for overtime.

Go/No-Go Decision

After completing the market and financial feasibility studies and presenting these to bankers and potential investors, the owners can make a good judgment as to the potential success of the food facility project. Further contacts with zoning boards, liquor license agencies, and other municipal groups will bring the project to a point of decision. The accumulation of the data contained in the feasibility studies together with encouragement or discouragement from lenders, investors, and municipal agencies will lead the owner to the first go/no-go decision. In other words, if the project looks financially sound, the market is identified, a need for the foodservice exists, and the capital is obtainable, the decision to go ahead can be made.

If one or more elements of the go/no-go decision are uncertain, there are three alternative courses to explore. The first is to correct the problem area that has been identified. The second option is to abandon the project and look for another place to invest the funds. The third alternative is to delay the decision until the final go/no-go decision point. This alternative is financially risky, because to progress from this point means incurring costs for foodservice facilities design consultants, architects, lawyers, accountants, and other professionals.

Site Selection

E. M. Statler's famous quote, "The three most important things for the success of a hotel are location, location, and location," is certainly true for many foodservice facilities. A poorly located restaurant will certainly experience a low level of sales, and a coffee shop in a hotel may miss a significant amount of business unless it has easy access both to the hotel guests and to street traffic. On a college campus, students will typically select convenience as the primary reason for eating in a particular food facility. Avoiding high rent by selecting a location that is inconvenient or out of the mainstream of foot or automobile traffic is usually a bad decision.

The location of a food facility on the immediate left or right of an entrance to a shopping mall is often considered a poor location, because the typical mall customer needs first to enter the mall and then look around for interesting places to shop and/or eat. Some considerations for site selection are:

— *Visual recognition*: Can the food facility be easily seen by potential customers? Will the appearance of the outside of the facility communicate the character and concept of the dining experience inside?

— *Convenience*: The developers of a destination restaurant, in which the customer has previously made a decision to dine, must consider parking, an attractive entrance, valet parking, and safe surroundings as important site selection criteria. Coffee shop, deli, and fast-food entrepreneurs must depend on impulse buying as the primary market, and therefore site selection must focus on signage, ease of entry, and drive-in windows and/or take-out service.

— *Code restrictions*: Site selection may involve local code requirements for setback from the street, parking capacity, street access, or the acceptability of a food facility in a particular neighborhood. Usually these code requirements are available from the engineering or city planning office of the municipality where the food facility will be located.

— *Environmental issues*: In suburban areas, the development of an attractive site for a foodservice facility may encroach upon wetlands or adversely affect water runoff in nearby neighborhoods. In older, urban areas, sites may have previously been used for industrial and manufacturing processes that left dangerous materials in the ground. Environmental factors such as these are increasingly important to the site selection process because they present additional costs.

Approval Process

The approval process for a new or renovated foodservice facility is typically long and complicated, involving many different agencies. Developers often experience the approval process-and the inevitable delays that occur-as frustrating because they are eager to get the facility open and operating so that it can generate a positive cash flow. In defense of those agencies and departments who must be satisfied before proceeding, it is worthwhile to recall that the project needs to be controlled so that it does not create an unsanitary, unsafe, or unattractive addition to the community. Each municipality has a different group of agencies who must approve the food facility, and each has a definite sequence in which the approvals are given.

For example, a health department may require preliminary approval of the equipment layout before construction begins and a final inspection after the facility is built and the equipment is in place. The way to obtain necessary approvals for a complex project is to develop a comprehensive checklist, in which each member of the planning team submits a list of necessary agencies and deadlines. After dates are recorded, one person (usually the architect) serves as the coordinator of the approval process. Typical approval agencies involved in foodservice projects include:

— Zoning board
— Health department
— Municipal engineers (water, sewer, gas, and electrical)
— City planner
— Fire marshal
— Liquor control board
— Telephone company
— State or federal agencies (on state or federal projects)

3

Food Processing

Serving good food is essential to the success of any restaurant or hotel, whether it offers exotic cuisine or hamburgers. Chefs, cooks, and food preparation workers are largely responsible for establishing and maintaining this reputation. They measure, mix, and cook ingredients according to recipes. In the course of their work they use a va]riety of pots, pans, cutlery, and other equipment, including ovens, broilers, grills, slicers, grinders, and blenders.

Chefs and head cooks often are responsible for directing the work of other kitchen workers, estimating food requirements, and ordering food supplies. Some chefs and head cooks also help plan meals and develop menus. Large eating places tend to have varied menus and kitchen staffs often include several chefs and cooks, sometimes called assistant or apprentice chefs and cooks, along with other less skilled kitchen workers. Each chef or cook usually has a special assignment and often a special job title-vegetable, fry, or sauce cook, for example. Executive chefs and head cooks coordinate the work of the kitchen staff and often direct the preparation of certain foods. They decide the size of servings, plan menus, and buy food supplies. Although the terms chef and cook still are used interchangeably, chefs tend to be more highly skilled and better trained than most cooks.

Due to their skillful preparation of traditional dishes and refreshing twists in creating new ones, many chefs have earned fame for both themselves and for the establishments where they work. The specific responsibilities of most cooks are determined by a number of factors, including the type of restaurant in which they work. Institution and cafeteria cooks, for example, work in the kitchens of schools, cafeterias, businesses, hospitals, and other institutions. For each meal, they prepare a large quantity of a limited number of entrees, vegetables, and desserts. Restaurant cooks usually prepare a wider selection of dishes, cooking most orders individually.

Shorat-order cooks prepare foods in restaurants and coffee shops that emphasize fast service. They grill and garnish hamburgers, prepare sandwiches, fry eggs, and cook french fries, often working on several orders at the same time. Fast food cooks prepare a limited selection of menu items in fast-food restaurants. They cook and package batches of food, such as hamburgers and fried chicken, which are prepared to order or kept warm until sold. Private household cooks plan and prepare meals, clean the kitchen, order groceries and supplies, and also may serve meals.

Other food preparation workers, under the direction of chefs and cooks, perform tasks requiring less skill. They weigh and measure ingredients, go after pots and pans, and stir and strain soups and sauces. These workers also clean, peel, and slice vegetables and fruits and make salads. They may cut and grind meats, poultry, and seafood in preparation for cooking. Their responsibilities also include cleaning work areas, equipment, utensils, dishes, and silverware. The number and types of workers employed in kitchens depends on the type of establishment. For example, fast-food establishments offer only a few items, which are prepared by fast-food cooks. Small, full-service restaurants offering casual dining often feature a limited number of easy-to-prepare items supplemented by short-order specialities and ready-made desserts. Typically, one cook prepares all the food with the help of a short-order cook and one or two other kitchen workers.

Working Conditions

Many restaurant and institutional kitchens have modern equipment, convenient work areas, and air conditioning, but many kitchens in older and smaller eating places are not as well equipped. Working conditions depend on the type and quantity of food being prepared and the local laws governing food service operations. Workers usually must withstand the pressure and strain of working in close quarters, standing for hours at a time, lifting heavy pots and kettles, and working near hot ovens and grills. Job hazards include slips and falls, cuts, and burns, but injuries are seldom serious. Work hours in restaurants may include early mornings, late evenings, holidays, and weekends. Work schedules of chefs, cooks and other kitchen workers in factory and school cafeterias may be more regular. Nearly 3 in 10 cooks and 1 out of 5 other kitchen and food preparation workers have part-time schedules, compared to 1 out of 7 workers throughout the economy.

The wide range in dining hours creates work opportunities attractive to homemakers, students, and other individuals seeking supplemental income. For example, about 27 percent of kitchen and food preparation workers are 16-19 years old. Kitchen workers employed by public and private schools may work during the school year only, usually for 9 or 10 months. Similarly, establishments at vacation resorts usually only offer seasonal employment.

Chefs, cooks and food preparation workers held more than 2.8 million jobs in 2000. Almost 60 percent of all chefs, cooks, and food preparation workers were employed in restaurants and other retail eating and drinking places. About 20 percent worked in institutions such as schools, universities, hospitals, and nursing homes. Grocery stores, hotels, and other organizations employed the remainder.

Training Cources

Most chefs, cooks, and food preparation workers start as fast-food or short-order cooks, or in other lower skilled kitchen positions. These positions require little education or training, and most skills are learned on the job. After acquiring some basic food handling, preparation, and cooking skills, these workers may be able to advance to an assistant cook position. Although a high school diploma is not required for beginning jobs, it is recommended for those planning a career as a cook or chef. High school or vocational school courses in business arithmetic and business administration are particularly helpful. Many school districts, in cooperation with State departments of education, provide on-the-job training and summer workshops for cafeteria kitchen workers with aspirations of becoming cooks.

Large corporations in the food service and hotel industries also offer paid internships and summer jobs, which can provide valuable experience. To achieve the level of skill required of an executive chef or cook in a fine restaurant, many years of training and experience are necessary. An increasing number of chefs and cooks obtain their training through high school, post-high school vocational programs, or 2- or 4-year colleges. Chefs and cooks also may be trained in apprenticeship programs offered by professional culinary institutes, industry associations, and trade unions. An example is the 3-year apprenticeship program administered by local chapters of the American Culinary Federation in cooperation with local employers and junior colleges or vocational education institutions. In addition, some large hotels and restaurants operate their own training programs for cooks and chefs.

People who have had courses in commercial food preparation may be able to start in a cook or chef job without having to spend time in a lower skilled kitchen job. Their education may give them an advantage when looking for jobs in better restaurants and hotels, where hiring standards often are high. Although some vocational programs in high schools offer training, employers usually prefer training given by trade schools, vocational centres, colleges, professional associations, or trade unions. Postsecondary courses range from a few months to 2 years or more and are open, in some cases, only to high school graduates. About 8 to 15 years as a cook is required to become a fully qualified chef. Those who gain experience, including in a supervisory capacity, may become executive chefs with responsibility for more than one kitchen.

The U.S. Armed Forces also are a good source of training and experience. Although curricula may vary, students in these programs usually spend most of their time learning to prepare food through actual practice. They learn to bake, broil, and otherwise prepare food, and to use and care for kitchen equipment. Training programs often include courses in menu planning, determination of portion size, food cost control, purchasing food supplies in quantity, selection and storage of food, and use of leftover food to minimize waste. Students also learn hotel and restaurant sanitation and public health rules for handling food. Training in supervisory and management skills sometimes is emphasized in courses offered by private vocational schools, professional associations, and university programs.

Across the Nation, a number of schools offer culinary courses. The American Culinary Federation has accredited over 100 training programs and a offers a number of apprenticeship programs around the country. Typical apprenticeships last three years and combine classroom and work experience. Accreditation is an indication that a culinary program meets recognized standards regarding course content, facilities, and quality of instruction. The American Culinary Federation also certifies pastry professionals, culinary educators, and chefs and cooks at the levels of cook, working chef, executive chef, and master chef. Certification standards are based primarily on experience and formal training.

Important characteristics for chefs, cooks, and food preparation workers include the ability to work as part of a team, a keen sense of taste and smell, and personal cleanliness. Most States require health certificates indicating that workers are free from communicable diseases. Advancement opportunities for chefs and cooks are better than for most other food and beverage preparation and service occupations. Many chefs and cooks acquire high-paying positions and new cooking skills by moving from one job to another. Besides culinary skills, advancement also depends on ability to supervise less skilled workers and limit food costs by minimizing waste and accurately anticipating the amount of perishable supplies needed.

Some chefs and cooks go into business as caterers or restaurant owners, while others become instructors in vocational programs in high schools, community colleges, or other academic institutions. A number of cooks and chefs advance to executive chef positions or supervisory or management positions, particularly in hotels, clubs, and larger, more elegant restaurants.

Job openings for chefs, cooks, and food preparation workers are expected to be plentiful through 2010. While job growth will create new positions, the overwhelming majority of job openings will stem from the need to replace workers who leave this large occupational group. Minimal educational and training requirements, combined with a large number of part-time positions, make employment as chefs, cooks, and food preparation workers attractive to people seeking a short-term source of income and a flexible schedule. In coming years, these workers will continue to transfer to other occupations or stop working to assume household responsibilities or to attend school full time, creating numerous openings for those entering the field.

Job openings stemming from replacement needs will be supplemented by new openings resulting from employment growth, as overall employment of chefs, cooks, and food preparation workers is expected to increase about as fast as the average for all occupations over the 2000-10 period. Employment growth will be spurred by increases in population, household income, and leisure time that will allow people to dine out and take vacations more often. In addition, growth in the number of two-income households will lead more families to opt for the convenience of dining out. Projected employment growth, however, varies by speciality.

Increases in the number of families and the more affluent, 55-andolder population will lead to more restaurants that offer table service and more varied menus-resulting in faster-than-average growth among higher-skilled restaurant cooks. As more Americans choose more full-service restaurants, employment of fast-food cooks is expected to decline and employment of short-order cooks, most of whom work in fast-food restaurants, is expected to grow more slowly than average. Duties of cooks in fast-food restaurants are limited; most workers are likely to be combined food preparation and serving workers, rather than fast-food cooks. In addition, fast-food restaurants increasingly offer healthier prepared foods, further reducing the need for cooks.

Employment of institution and cafeteria chefs and cooks also will grow more slowly than the average for all occupations. Their employment will not keep pace with the rapid growth in the educational and health services industries-where their employment is concentrated. In an effort to make "institutional food" more attractive to students, staff, visitors, and patients, high schools and hospitals increasingly contract out their food services. Many of the contracted food service companies emphasize simple menu items and employ short-order cooks, instead of institution and cafeteria cooks, reducing the demand for these workers.

Food Producing Companies

There are three areas of food producing companies that somehow have a relation to planning will be described (Figure1). First, describe typical market characteristics of food processing industries. Second, some organizational characteristics and typical organizational structures will be discussed, and third, describe the production system as it appears in most small and medium sized food processing industries.

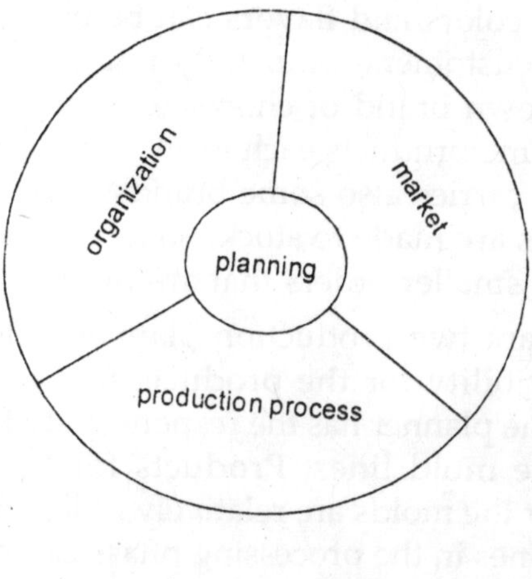

Figure 1. Elements that relate to planning

Chocolate Factory

A manufacturer makes chocolate in all kinds of forms such as bars, sticks, and coins. The production process of all the different products is roughly as follows. First, the liquid chocolate that is stored in tanks is transported to the appropriate mixer. A human operator puts ingredients such as sugar, butter, flavors, and coloring matter to the chocolate, and the substance is mixed for a while. The substance that is now available in the mixing tank is one batch. The mixing tank can not be used again until it is empty. From here, the process is a continuous flow until the products are finished. The substance is first put through coolers which is a delicate process with very small margins for the speed of cooling.

After that, the cooled chocolate is put through extruders or in molds. Two production lines use an extruder, and two production lines use a mold. Some additional operations must be performed, such as cutting, sorting, putting the products in trays, wrapping it in tinfoil, and putting it in boxes. All coolers are connected to all extruders and molds, which means that the variety that is available in colors and flavors can be used for all product forms. The customers are mainly retail organizations that carry their own brand of chocolate bars. These customers order the same products each week but the amounts differ. The factory carries also some brands of its own for which the products are made to stock. Some large orders are split into several smaller orders that are produced separately.

There are two production planners. One planner has the responsibility for the products that use the extruder lines, and one planner has the responsibility for the products that use the mold lines. Products for the extruder and products for the molds are relatively independent, although some machines in the processing phase can be used by both and some machines in the packaging phase can be moved from one line to another.

The planning of the molded chocolate is mostly for the Dutch market and around 90% is produced to stock. The main goal is to keep the stock levels high enough. As a rule, stock levels should be three weeks. Customer orders have

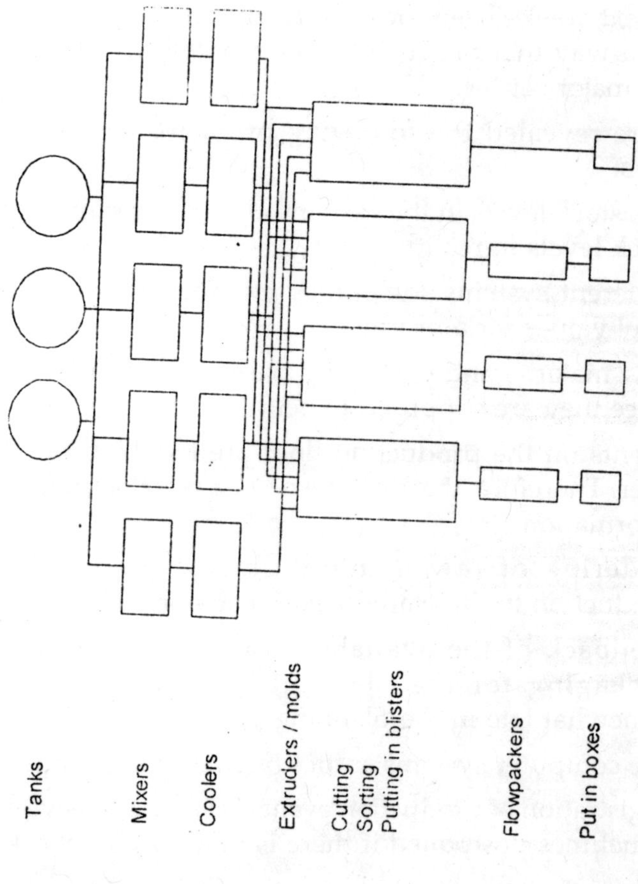

Figure 2. Layout of chocolate factory

a delivery time of 10 working days (2 weeks). The extruded chocolate is mainly sold abroad. About 90% is produced to order and only a small amount is produced to stock. The planning process is roughly the same for both planners. First, at the beginning of the week a list is made that contains customer and stock orders that must be produced in the next week. Then, orders are clustered and ordered in such a way that production can proceed for a full day without major setups.

Here revealed the following problems or potential problems:

— The stock levels in the systems are not the same as the stock levels in reality.

— Different systems sometimes provide different stock level values for a product.

— Raw material and packaging are sometimes not at the place they are supposed to be.

— Events on the production floor are fed back the day after. Therefore, there can be no fast reaction to formal information.

— Ordering of raw material and packaging from production to the warehouse is sometimes late.

— Feedback of the availability of raw material and packaging for the planning (the simulation) is somewhat late in the planning process.

— The computer systems suffice but are rather outdated.

— Registration of production events is labor intensive and sometimes postponed if there is more urgent work to do.

— Alternative routings for products are not structurally known.

— Delivery performance is not as high as the manufacturer wants.

All in all, the factory appears very well organized, and these problems only indicate possibilities for improvement.

Cookie Factory

A manufacturer of filled wafers has the following production process. First, large wafers are made in ovens. These wafers can be used immediately, but they can also be put in a heated room. If the wafers are not used immediately, they attract moisture, which causes the wafers to get out of shape. Hence, if they are put in the heated room, they must lay there for at least two weeks, after which they are straight again because they have dried fully by then. The next processing step is filling the wafers. This means that crème is put on one wafer, after which another wafer is put on top of it. This can be done for a maximum of nine levels. Subsequently, the wafers are cut and sometimes topped with chocolate. After this, the wafers can be put into a tray, and the trays are flow-packed and boxed. The packaging form differs per product (Figure 3).

Packaging machines can be moved to form a packaging line. There are six possible packaging line configurations but at most four can be run simultaneously due to scarcity of packaging machines. Baking the wafers, filling and cooling the wafers, and cutting and packaging the wafers are decoupled from each other. Both baked wafers and filled wafers can be stored.

Customers are retail organizations both in the Netherlands (around 40%) and abroad (around 60%). The number of customer orders for the Netherlands is around 40 per week. Around 75% of the local orders are delivered from stock, and 80% of the Dutch customers have a fixed delivery day each week. There are approximately 5 foreign orders per week that are each composed of 2 to 4 different products. The size of foreign orders is typically a full truckload or container, which is far larger than domestic orders.

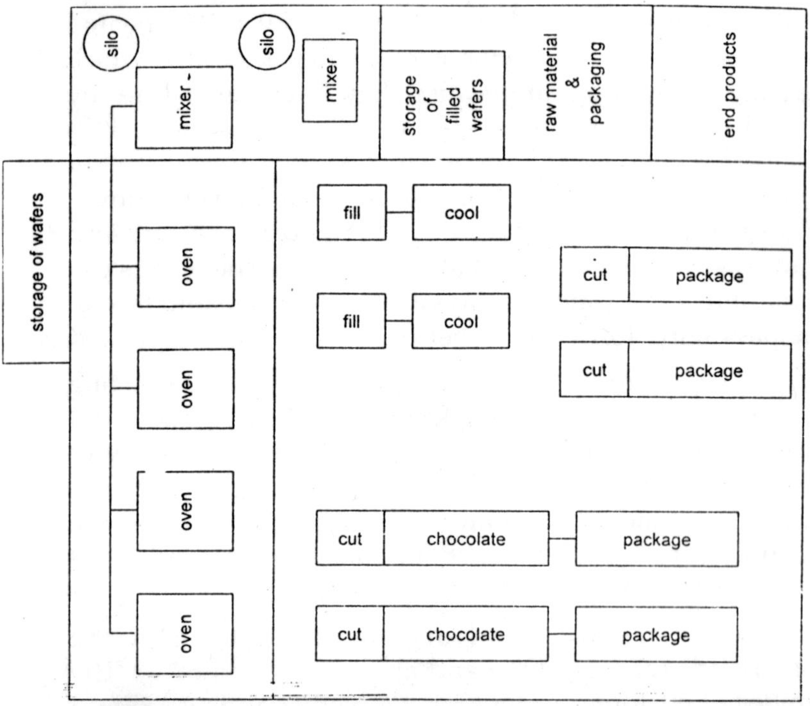

Figure 3. Layout of cookie factory

The production planning is split in four stages:

— the planning of the orders that must be produced,
— the planning of the packaging lines,
— the planning of the fill lines, and
— the planning of the ovens.

The finished order planning is the basis of the plans for the packaging lines, the fill lines, and the ovens. These production stages are decoupled in the production process, so the plans can be made somewhat independent. The preferred sequences of production differ in the subsequent stages. The plan for the packaging lines is made once a week, and the plans for the fill lines and the ovens are made on a daily basis. However, the order planning and the packaging line planning are adjusted frequently during the week. The company uses a system for logistical control that handles order processing, accounting, and planning. There is also a spreadsheet that contains the recipes and that can calculate the needs for raw material and packaging on the basis of the amount of end products that must be made.

There are several problems that make the production control system somewhat inappropriate. First, the system does not account for capacity. The order planning module only accounts for due date and throughput time. Second, the time of operations is rounded up to at least one day. Therefore, it is not possible to plan multiple operations on one day. Third, operations and materials are planned per week, so it is not possible to plan each day for the following day. Here concluded the following problems or potential problems:

— There are four people that are each responsible for a part of the planning. Because the plans are dependent, those four people must meet and discuss their plans regularly. This makes the throughput time of the making of the plan rather long. Furthermore, the way in which changes of one of the plans is communicated is mostly ad hoc. Therefore, it can happen that someone learns about a change too late.

— Raw material and packaging is sometimes not available. The cause is that it is checked somewhat late if everything that is required according to the plan is actually available. The place that needs the plan as the first is the fill lines, but it gets the planning as the latest.

— Computer support for the planning is mostly absent. The planning is made manually on paper or in a word processor.

Although the case descriptions are of course examples, we have seen these kinds of problems occur in other companies as well.

Character of Market

The value chain of the food industries proceeds from farmer to industry to retailers to consumers. Meulenberg & Viaene state three reasons why food retail chains have substantial bargaining power. First, food retail chains have concentrated to an oligopolistic market. Macchietto reports about this that "the industry is characterized by a small number of very large players, with global, integrated product development, manufacturing and distribution networks, and strong product brands.

In the UK 5 big retailers share approximately 80% of the market. They have very large buying muscles and command much attention from their manufacturing suppliers. The retailing end of the business is characterized by high volume, low inventory, short lead times, wide variety of products, and strong competition on price". Second, there is a surplus of production capacity in food processing industries. Third, products are often only to a limited amount unique and producers are exchangeable.

One of the effects of the changes in the power balance is that the retail organizations can require that the products can be replenished at short notice. Or, reasoned from the perspective of the producer, that the order lead time must become shorter (Table 2). This shortening of lead times has consequences for many aspects that are engaged in production, such as order processing, materials management, planning, and shipping.

Table 1. Developments in the food supply chain

Stage in the supply chain	Developments
Growers/ producers	Increasing production costs due to governmental rules concerning environmental and consumer-related issues
	Lower prices due to liberalization of markets
	Reducing number of and scaling-up of farms in the EU
Wholesalers	Scaling-up and concentration
	Global sourcing
Food industry	World-wide concentration of food producers
	Increasing power of retailers
	Differentiation by A-brands
	Advanced processing and Information and Communication Technologies
Retailers	World-wide concentration of retailers
	Growing strength of supermarket own-label products
	More consumer knowledge through new Information Technologies
	Growing relative importance of supermarkets for grocery purchase
	New ways of distributing food to consumers
Consumer markets	Saturated markets
	Mass customization

Table 2. Lead-time from supplier to distribution centre (DC) to store at Albert Heijn.

Lead-time	Supplier/DC	DC/Store
Past	120-48 hours	48-36 hours
Present	48-24 hours	18-12 hours
Future	12-4 hours	18-4 hours

Demographic changes in the population of consumers and changes in consumer behavior have also resulted in changes in food processing industries. Families become smaller and the number of two income households becomes larger. Heilig discusses three trends. First, traditional dishes that are prepared from raw products are being replaced by industrially produced food. Second, seasonal patterns in food consumption disappear, and third, exotic food is gaining popularity. Other trends are that life-span of products is shortened and that there is a tendency of diversification.

Character of Organization

The number of employees of small and middle sized food processing industries typically ranges between 30 and 150 for the production floor. Production mostly stops at nights and in weekends but sometimes production also proceeds then. The organizational structure depends mostly on the history and ownership of the firms. Most of these companies have an origin as a family business that has grown throughout the years. This has consequences for the way that such organizations are managed. Decisions in firms that are owned by the managing director are usually taken in a more centralized way than in firms that are a division of a large corporation. There is a trend of concentration in the sfood processing industries, although not as extreme as in retail.

Mintzberg distinguishes the strategic apex, the middle line, the technostructure, the support staff, and the operating core (Figure 4). The strategic apex generally consists of the board of directors, led by the managing director. Common functions in the board of directors are a marketing manager, a financial or accounting manager, and a plant or production manager. Depending on the size and complexity of the firm, there could also be managers for research and development, logistics, and other functional areas. Purchasing of critical raw material that needs to be bought with long term contracts might also be a task of the board of directors.

Figure 4. Organizational structure in food processing industries.

The operating core deals with the primary process. Functional areas with respect to the flow and transformation of goods are purchasing, warehousing, production, sales, and shipping. The support staff consists of functions such as legal counseling, public relations, research and development, quality control, etc. Finally, the technostructure's purpose is to make the work of others more effective. Here, apparent functions are strategic planning, training, maintenance, process control, and production planning. Due to the size of many companies in the food processing industries, employees often perform multiple functions, or functions are not performed at all. Formal authority relations and coordination of work are mostly organized by process and function. This means that work is organized in a traditional hierarchical way.

Planning is often the link between marketing or sales and production. Therefore, the placement in the hierarchy has an impact on the power balance. The appropriate place of production planning in the hierarchy is, both in organization theory and in practice, not clear. In most cases, the planning department is positioned in one of four possible settings: there is a planning department next to the production and sales departments; the whole planning department falls under the production manager; the whole planning department falls under the sales manager; or some people in the planning department fall under the production manager and some people fall under the responsibility of the sales manager.

If the planning is positioned either in production or sales, there can be a clear emphasis of interests. In the mixed form, some people in the planning department function under the production department, and some people function under the sales department. Then, a conflict in interests can first be tackled in the planning department itself, before higher levels in the hierarchy must be involved. Even if the organizational structure and authority relations are agreed upon in an organization, perceived goal differences between employees of marketing, production, and planning can impede the "functioning of organizations as single integrated wholes".

Character of Production System

The production system consists of all activities that engage in the production of food. These activities are not limited to physical transformations. Activities such as internal transportation and some management activities also belong to the production system. In this section, we describe two parts the elements of the production system that usually exist in food processing organizations.

Physical Aspects

As said, a production system is a collection of interrelated activities involved in the production of goods and services. The elements of a production system can be modeled as consisting of inputs, conversion and creation processes, and outputs (Figure 5).

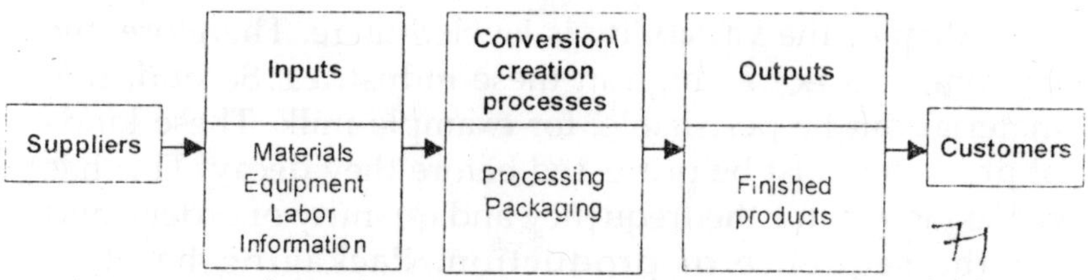

Figure 5. Production system

Materials

Two kinds of materials can be distinguished: raw material and packaging. Raw material is used to actually make the products. The raw material suppliers for consumer food producing organizations are mainly industrial organizations that make bulk products. Usually, there are two kinds of raw material for consumer food producers. First, there is raw material that constitutes most of the product and that is used in high volumes. Examples are liquid chocolate, flour, butter, milk, sugar, water, etc. These kinds of products can usually be delivered at short notice and in high volume. Second, there is raw material that is used in smaller quantities, for example flavors, scents, coloring matter, etc. These substances are often predominant at distinguishing the product features from a customer point of view.

Procurement lead times of these products can be long. Raw material in food processing industries has two important features that influence production control. First, the agricultural nature makes that variability in certain aspects is high. It is not always exactly known beforehand how much will be delivered, the quality of the raw material can vary, and the price of raw material can fluctuate, e. g., with seasons. This variation can lead to variations in the recipe to keep the product characteristics constant. Food processing industries that make consumer products usually get their raw material from food industries that (pre)process agricultural material.

Most of the variability is leveled there. Therefore, the variability is not as high in these industries. Second, raw material can be perishable, for example milk. These kinds of products must be processed before they decay. This has consequences for the frequency and quantity of orders, and for the sequence of production. Packaging has two functions. First, it is used to protect the food from damage and decay. Second, packaging determines the appearance of the product. Much products use both kinds: an inner wrapper for protection, and an outer wrapper with brand specific depictions and product information. Often, packaging is available on demand. Then, the organization itself has a only a small stock of packaging, and the supplier can replenish it at short notice.

Functions of Equipment

There are two types of functions of equipment in food

processing industries: processing and packaging. These two functions are mostly performed by different machines, but there are also machines that can perform both functions. In the remainder, the term equipment will be used to refer to a machine from a functional perspective. Processing equipment transforms raw material in the eatable or drinkable product. Packaging equipment puts this in packaging. The differences in these kinds of equipment lead to a separation in a processing phase and a packaging phase in production. In the processing phase, the equipment typically consists of kettles, ovens, pipelines, mixers, conveyor belts, etc.

In the packaging phase, packaging machines are used to put foil or tin foil around the product and to put the discrete units into trays, tin cans, or boxes. Somewhere in the processing phase or the packaging phase, the products must be transformed into discrete units. From a production control perspective, the technical aspects of such equipment is not important. Rather, it is functional aspects that can influence planning decisions:

— One batch in the processing phase can be used for multiple products in the packaging phase. Therefore, orders that can be created from one batch should be planned together to fill one such batch.
— Buffers make that processing can be planned separately from packaging.
— Change-over time (cleaning of and changes to equipment due to product changeovers) during production hours means a loss of production time. Planning must take this into account because change-over time can be sequence dependent (e. g., a pipeline must be cleaned if white yoghurt is produced after strawberry yoghurt, but not the other way around).
— Differences in production speed can indicate preferred machines for some products.

The layout in food processing industries is mostly a mix of process layout and product layout (Figure 6).

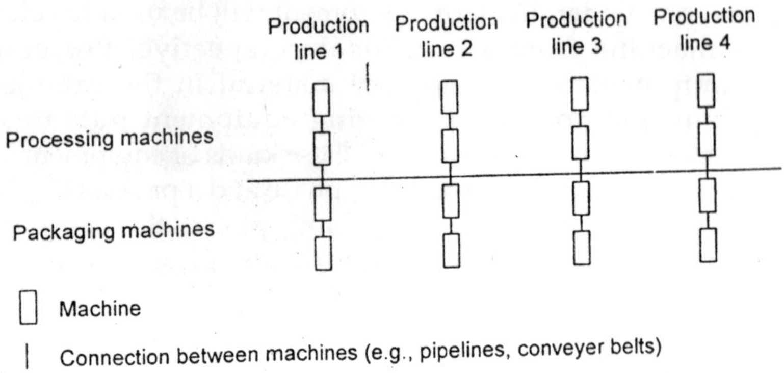

Figure 6. Layout of food processing industries

Because all processing machines are grouped together and all packaging machines are grouped together, there is a process layout. But, since the machines are connected in the sequence of processes, there also is a product layout with batch production. Note that there can be the possibility to change the flow, i. e., a product that is processed at production line 1 can be packaged at packaging line 3. The occurrence of both process and product layout induces specific issues for coordination at the production floor, which will be discussed in the subsection about control aspects.

Information

Both the information content and information flows depend heavily on case specific characteristics, but there are several generic information types that are used in the production system. First, personnel must know what must be made at what time. This information comes from the planning. Often, the planning only gives a sequence of orders (sometimes with due dates) without specifying exact starting and ending times. Second, it must be known how to make the products. This information is often automated, but books in which operators must manually search recipes are also frequently used. Third, there must be coordination between the several phases in the process, e. g., the packaging department must know when exactly one batch will be finished and the next will be started to be able to prepare a setup. Products that are discrete can be identified by physically attaching product information, something which is more difficult with non-discrete products. Material is transformed by equipment on the basis of information. Here elaborating production processes in food processing industries.

Food Processing

Although food is usually composed of several ingredients, production is mostly divergent because of a large variety of products that is created by different packaging forms, and sometimes by-products. Some processing operations can occur after packaging, e. g., cooking of sausages. The processing phase has characteristics of both continuous and batch production. The packaging phase is often purely batch oriented. The throughput time of a batch in both kinds of processes is often small, in the order of magnitude of hours, although processing steps can sometimes also take longer, e. g., days. Following are the both kinds of processes.

Processing phase: In the processing phase, the actual food products are made. This is done by all kinds of physical, chemical, and biological processes, e. g. :

- Separation processes (e. g., diffusional extraction, concentration of juices, separation of flour)
- Assembly or texturizing processes (e. g., emulsion processes for margarine and ice cream, foaming of whipped cream, extrusion processes, dough making)
- Bioconversion processes (e. g., sugar fermentation to alcohols, roasting)

— Preservation processes (e. g., retorting to eliminate microbial, enzymatic or chemical spoilage).

Processes in the processing phase are often variable to a certain extent. The duration and yield of processes can both fluctuate. This uncertainty can result in control problems. Ways to deal with this are to use a mean yield in the recipe, create a safety buffer of raw materials that have the most variable yield, create safety time in the production plan, and to maintain a stock of end-products. Sometimes, there is a pre-processing phase (to make some half-products beforehand) that can be performed independently of the main production process because the half-products can be stocked. This is important for production control because it allows decoupling of processes.

The processing phase has some characteristics that are associated with continuous or flow production because it deals with bulk quantities of homogeneous products. On the other hand, there are several reasons that operations that are basically continuous in nature are nonetheless produced in batches. First, there is only a limited amount needed of a product. A following product might need another recipe which results in a forced interruption of a continuous process. Second, some processes require batch-wise production because operations such as mixing and heating take place in kettles. Third, hygienic requirements necessitate frequent cleaning.

Processes that are continuous need periodical interruptions for cleaning, and cleaning operations need to be scheduled. Fourth, product decay leads to a limitation of storage life which in turn can lead to separation into batches. Fifth, lot traceability requirements can lead to artificial separation into batches. The charge unit size is often a tank or a vessel, and the batch size is a number of charges. From one batch, a lot of consumer units can be produced. Often, even several types of consumer products can be made from one charge by switching the packaging size or brand. An effect of production in batches is that if quality is poor, the whole batch must be rejected.

Packaging Phase: Two functions of packaging can be distinguished: protective packaging and cosmetic packaging. Some packaging has both functions. Examples are cartons or bottles for dairy products, and trays with an airtight plastic wrapper for biscuits and cookies. Similar products can be packaged differently. Examples are differences in packaging size, and differences in the labels on the packaging. According to our observations, production interruptions occur more often in the packaging phase than in the processing phase, especially if the products are small and discrete, e. g. cookies. Products sometimes fall crooked on conveyor belts, thereby causing congestion.

In addition, it is not rare that machines are old and frequently get jammed with the product or with the packaging. The packaging phase is usually labor intensive. Sometimes parts of the packaging phase are entirely manual, for example removing broken products, putting biscuits in trays, etc. Automation of such processes is difficult and expensive. In addition, somebody must monitor the automated processes. Because a line-stop can result in losses of products, products are manually put on tables as much as possible if a packaging machine is jammed. An important aspect in practice is that packaging lines need some time before they run smoothly after a major setup, for example because the final adjustment of the setup can only take place during running, or the line operators must get used to the current product. This is one of the reasons why frequent setups are avoided.

Outputs: The output of production processes can be separated in four categories: semifinished or half products, end-products, by-products, and waste. Half products are usually stored only shortly at the production floor because shelf life and storage space are limited. The end-products are mostly stored on pallets to be shipped to a distribution center at short notice. By-products are often sold, and waste is thrown away, sold to for example feed industry, or stored to be reused in future products. The description of inputs, process, and outputs concludes the discussion about the physical aspects of the production system.

Production Control

Production control deals with the control of transformation of goods or services in time. Such functions deal, for example, with the flow of materials, utilization of people and equipment, and coordination of internal activities with suppliers. Production control decisions must balance the requirements of the market, such as delivery performance and product quality, with internal organizational factors, such as working hours, setup costs, production batch sizes, capacity usage, etc. Here distinguish two main categories of production control decisions. One is decisions with respect to materials management and another is decisions with respect to capacity use.

Materials Management: There are several aspects of materials management that are common for small and medium sized food processing industries. First, material management mainly deals with stock management of raw material, packaging material, and finished goods. Material flows in food processing industries, which can be very complicated in a job shop, do not need special management due to the flow characteristic of facility layout and relatively short throughput times. The short throughput times also lead to a small amount of work in process compared to the stock of raw material and finished goods. Second, stock of finished goods is usually not stored very long at the production facility.

The purchasing function decides what goods will be ordered and the moment to order them. This function must work closely together with both production control to ensure that the right products are ordered timely, and with inventory management to ensure that the stock positions will not become too high or too low. The warehouse is responsible for storage of goods that are delivered by the supplier and products that are made by the production department. Often, the warehouse is also responsible for the transport of those goods to the production floor and back. In food processing industries, there are two main aspects that determine how the goods will be stored.

First, volume can be high or low. Second, the type of product can be discrete or nondiscrete. High volume non discrete products are usually stored in silo's and tanks. These are not necessarily placed in the warehouse. Rather, the location depends on the way they must be filled or emptied. Transport of such products mostly takes place by pipelines. Inventory management functions must decide on the stock positions for the various products. Some aspects that must be taken into account are costs, decay, order lead time, and the expected use of material.

In small and medium sized food processing industries, inventory management functions are usually performed by the purchasing department, the production manager, the planner, or the foremen. Stock positions can be calculated by production norms (current stock is the previous stock minus what is used for production according to the recipe) or by measurement. Although the latter is more accurate, the former is used more often because it takes less time and is accurate enough in most cases. There is, however, some variability in what is actually used during production. Frequent stock measurement is then used to make sure the stock figures do not deviate too much from reality.

Production Planning

The combination of a product layout and a process layout has some consequences for production control. First, work at the functional areas must be coordinated. For example, if only one person must monitor all kettle processes, then there should not start too many new processes at the same time. Second, work at the production lines should be coordinated. This has mainly to do with the moment of product changeovers and production speed of the machines in the line. Such decisions are made at the production floor. There is usually almost continuous mutual adjustment about production speed and the exact moment of product changeovers. An overall coordination mechanism in the form of a production plan states what products must be made at approximately what moment.

With the information from the production plan, all departments know what activities must be employed. It is necessary to know this in advance because there is often the need to make some preparations, to order raw material and packaging, and to make a personnel schedule. The production plan is made on the basis of expected and received customer orders and stock levels, within the scope of available raw material, available packaging, and capacity restrictions. The production plan usually states what products should be made in what time frame, and is based on average production speed. There exists several production control frameworks in which planning activities are described. The frameworks include activities such as resource planning, demand management, master production scheduling, capacity planning, and materials planning.

Example frameworks are MRP-II for discrete environments, and the Process Flow Scheduling framework for process industries. The frameworks have separate sub-plans in an aggregation sense as well as in a decomposition sense. Production planning and control frameworks scarcely pay attention to day-to-day planning. The models in the frameworks usually stop at the level of order release, i. e., an order is released to the factory with a due date before which it must be produced. When the order must be handled exactly is then yet to be determined. How this must be decided on is usually not prescribed or described by production control frameworks.

The actual assignment of orders to machines and production times is what we have called the day-to-day planning task. The path that leads to such a plan usually contains several levels. An apparent distinction in levels that stems from Operations Research is the one between sequencing and scheduling. Sequencing decides on the number and order of batches of different products that need to be manufactured to fulfill given production requirements.

Scheduling determines the precise timing of each batch of material going through the system. Van Dam and Van Dam et al. describe two levels with day-today planning activities they encountered in practice: a higher level labeled the Detail Planning Level (DPL), and a lower level labeled the Scheduling Level (SL)

The DPL deals with decisions that have an important impact on other scheduling decisions. DPL decisions often establish prerequisites that extend over individual orders, i. e., decisions at the detail planning level are made on the basis of aggregate measures or forecasted information. Decisions at the DPL usually concern balancing of capacities for the coming weeks, decisions with regard to overtime, number of shifts, number of employees, stock levels, purchasing of materials, the moments of major product changeovers, possibly the sequence of production, etc. These decisions determine the decision space for the scheduling level. Decisions at the SL can deal with allocation of production orders to production lines, assignment of operators to production lines, sequencing of orders, starting and finishing times of production, etc.

There are two kinds of decisions at the SL. First, constructive scheduling decisions result in a schedule of a period before that period starts. Second, reactive planning (also called repair-based planning) means that an existing plan is adapted. Repair-based scheduling decisions are based on unexpected events that occur during the period that the plan covers. Such events are, for example, stock-outs, machine breakdowns, rush orders, etc (Table 3). The division in a constructive phase and a reactive phase is also called offline scheduling versus online scheduling. There are often three parties involved in the planning process: the production manager, one or more planners, and one or more foremen or operators in the production process. The allocation of decisions is highly case dependent, but some general remarks can be made. The production manager often deals with some decisions at the DPL that have an important impact on costs, such as the number of shifts, the production lines that are going to be used, the number of setups, etc.

Table 3. Examples of events

Source	Event
Customer	New order
	Change of order
	Cancellation of order
Product	Rejection of batch due to quality problems
	Raw material out of stock
	Packaging out of stock
Process	Process duration differs from planned (e. g., due to short machine interruptions or processing uncertainties)
Machines	Breakdowns for longer periods
	New capacity available

The planner deals with decisions at the DPL that have a more operational nature, such as determination of the size of stock orders. In addition, the planner makes SL decisions such as allocation of orders to production lines, sequencing decisions, and rescheduling. The foreman can sometimes decide on the sequence of production, and the exact starting time of production of an order.

Analysis of Conflict

The design of a planning concept in production organizations must weigh the interests of the production department, the interests of the customers, and the efforts that are needed to make the plan. The following overview describes how a higher order frequency, smaller orders, and shorter lead times contradict the effects of the production characteristics.

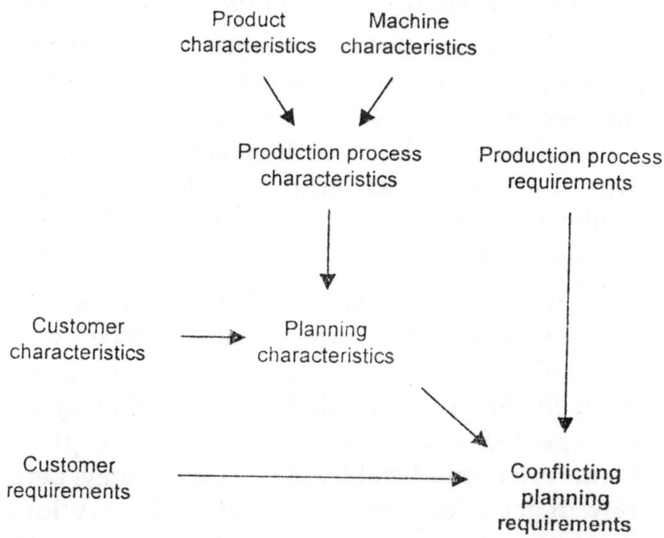

Figure 7. Conflicting planning requirements

— *Small orders* : Due to setup-times, cleaning times, and startup losses, long runs are preferrable from a production perspective. Production batches are often preferred to provide a half day or a whole day of production time so changeovers can take place during lunch or in the evening. This contradicts the wishes from the market that orders should be small. Customer orders can be grouped, but this contradicts the requirement of short lead times, as will be explained next.

— *Short lead times and multiple products from one batch* : Sometimes several orders can be produced from one processing batch by changing the packaging. To change the packaging is usually not very time consuming. Therefore, grouping can be used to process small orders efficiently. However, because the processing batches (which are the groups of customer orders) must be attuned, the customer orders must be known beforehand. This opposes short lead time

— *Short lead times versus sequence dependent setup and cleaning times* : Due to sequence dependent setup-times and sequence dependent cleaning times, orders should be known beforehand so they can be sequenced in a way that minimizes setup and cleaning losses. This contradicts the wishes from the market that lead times should be short.

— *Short lead times and low inventories* : If the lead time of customer orders is shorter than the procurement lead time, then raw material and packaging must be stocked. Stock is avoided for a number of reasons: unexpected stock-outs can occur if the processing yields are unpredictable, stock costs money, perishable raw material can not be put to stock very long, and producing from stock means that the manufacturer must sometimes reject an order if there is not enough.

— *Short lead times and variability in processes* : The focus on high use of capacity usually leads to tight schedules in which there is not much slack to deal with variability in the production processes. Production to order with short lead times and low stocks of finished goods enlarge the risk that variability leads to stockouts or tardiness.

— *Short lead times and planning organization* : The organization of the planning is usually not geared towards adaptation of the plan. If lead times are shorter than the planning horizon, however, then orders should be processed in the existing schedule. Due to the sequential nature of much planning decisions, it proves difficult to adjust an existing plan.

Short lead times and small order sizes clearly oppose efficient production in a number of ways. Most of these contradictions have their origin in physical aspects. Reducing cleaning times, setup times, startup losses, and variability of production processes, would remove a lot of barriers to short lead times and small orders. The contradiction that refers to the planning organization is not of a physical nature but has to do with the control of the production system. If plans could be adjusted easily, then lead times could be short, and physical characteristics would be the bottleneck instead of the planning.

4

Food Planning and Management

Many foodservice operations are family-owned or individual-owned enterprises, but this characteristic has been constantly eroding over the last thirty years. Today, massive foodservice chains do most of the dollar business, and the small, single-outlet enterprise's share of the market is shrinking. The large corporate organization is the leader in the industry. Within the last thirty years, the type of food used in foodservices has gradually changed. More processed foods have found their way into use, some of which are practically ready to serve. Many operations still want to produce their own food, but even they now use more processed foods to do it. Another development is the central commissary, where foods are mass-produced and then shipped to satellite units to be served.

The problem of maintaining quality while obtaining desirable economies has slowed the growth of central commissary production, however, and some operations that had instituted central commissaries have now discontinued them. The industry has had to change to meet a changing patron market, too. People eat less, have clearer concerns for their health, are eating out more, and are more discriminating shoppers. The industry has become highly competitive, and its management is now largely composed of people who have had training in business or in specialized schools that teach foodservice management.

Employers in Foodservice Industry

The foodservice industry employs millions of people, and each year several hundred thousand new workers must be found to fill newly created or newly vacated jobs. Until recently, much of this new labor was recruited from "off the streets." These people simply walked in and were assigned jobs whose duties they learned by experience. Their teachers were other workers who may or may not have known how to do the job; as a result, the new worker often learned to do the job incorrectly or inefficiently. Upperechelon jobs were often filled by promotion from within the operation's existing ranks.

Not infrequently, the manager or even the president of a large corporation was once a pot washer or bus person. Within the last forty years, a sizable educational system for training employees at all levels of the industry has emerged. Unions, too, have concerned themselves with educational courses-in part so that they can certify the job abilities of members better. Unions often work with public schools, but in some cases they have established their own training and certification centers. The government has also been active in promoting short educational programs to train workers for the foodservice industry.

Professional associations have set up programs to give educational opportunities to their members. The Club Managers Association has developed an educational program that must be taken by any member who wishes to use the title CM (Certified Manager). The American Dietetic Association requires all dietitians to be certified by examination and to continue to accumulate educational and professional points during their professional lives if they intend to use the title RD (Registered Dietitian) after their names. The School Foodservice Association and some of its state organizations now require all workers to gain educational credits in order to be certified.

Many universities and other schools sponsor training programs for workers, supervisors, and managers. At some institutions, 5000 or more people per year register for correspondence courses. Many high schools offer courses for waiters, waitresses, cooks, and other service positions. Many community colleges and junior colleges have excellent programs for training workers. The emphasis varies from vocational courses for those wanting to work as chefs or maître d's to economics and management training for people interested in becoming food and beverage managers, sales managers, or catering or banquet managers. Four-year hospitality programs commonly accept credits from these schools. Nearly 200 universities and colleges have four-year programs in the area, and some offer master's degrees and a few the doctorate. Over 2000 graduates from four-year schools are thought to enter the foodservice field each year-and the industry could use twice that number.

Each year companies make recruiting visits to educational institutions, offering graduates a three month or longer training program before being moved into supervisory or managerial positions. Because of the rapid growth of many businesses, some graduates hold important positions within four or five years after leaving college. The growth of graduate programs to develop top executives and teachers for the industry has also been significant. Most four-year programs are offered in the College of Business, but others may be found in the College of Home Economics or the College of Food Technology, and still others exist as separate schools. The discipline of foodservice management is today well recognized in academic circles.

Most academic programs require that students spend many hours working in the industry; internships have been established whereby students gain on-the-job experience and receive academic credit. Such an approach works best because-although many basic facts can be learned in class-the student must be exposed to situations that are related to the real work world·in order to develop into a successful manager. Our workforce in the foodservice industry is rapidly changing. Native Americans, Latinos, Asians, and African-Americans now make up a significant part of the labor force, requiring management to understand cultural differences among employees. Also, the number of older people and those with disabilities in the foodservice industry has increased significantly.

Workforce Diversity

Workforce diversity is no longer just a gender or religion management consideration. While it originally referred simply to a person's gender or ethnic group, today, workforce diversity encompasses differences in age, affectional orientation or sexual preference, economic status, educational background, ethnicity, gender, geographic location, income, marital status, lifestyle, and religion, to name a few characteristics. Foodservices that manage workforce diversity efficiently benefit from more effective recruiting, improved management quality, increased market share and productivity, and reduced labor costs through lower employee turnover.

Although developing and maintaining programs that enhance diversity can be difficult to implement, the benefits far outweigh the disadvantages. A comprehensive and carefully planned approach for implementation is required. One foodservice company, McDonald's, has been successful through its Changing Workforce Programs, whereby different courses are offered to help employees work through gender and cultural differences. These courses, such as Managing the Changing Workforce, Women's Career Development, Black Career Development, Hispanic Career Development, and Managing Cultural Differences and Managing Diversity, encourage open communications.

The end result is a raising of the standard of professionalism within the foodservice industry. As labor is tight in many markets, workforce diversity must play an even greater role in the success of foodservice operations tomorrow. At the present time, Latinos and African-Americans make up 25 percent of our total population, and this percentage is expected to grow.

Productivity of Employee

Another concept that raises the standard within the industry is employee productivity. People who enjoy coming to work generally produce more, as will workers who believe that their work will be valued or lead to advancement. The challenge is to clearly define what good work is for the employee and how that can be measured. Reward systems are the key to recognizing individual contributions in order to encourage greater productivity. In one operation, ability to work in any station within the kitchen was looked upon as excellent work. That way, employees could assist others in accomplishing tasks at hand when a rush would come.

Accordingly, employees were encouraged to learn each other's job and were recognized as having done so by receiving a pin to wear for each station learned. With the knowledge of sauté, broiler, saucier, garde-manger, and other positions under one's belt, that employee became highly productive for the kitchen as a whole. Morale improved and turnover was lowered. With improved productivity, any foodservice operation will attract and retain employees better. A positive atmosphere in which employees are recognized for their contributions, rather than being criticized for their mistakes, will do much to improve productivity.

Supervisors should look for employees to do the job approximately correctly, at which point recognition should be given. Food production issues are viewed from the standpoint of what a manager ought to know to be able to manage these operations well and to ensure that adequate standards in food production and service are satisfied. The management skills needed to operate a foodservice are presented in such a way as to prepare the reader for the responsibilities of an entry-level position in the industry.

Management is sometimes defined as the element in an enterprise that plans, staffs, organizes, directs, and controls an enterprise so that it reaches its goals. It is management's job to establish goals, plan how to reach them, and then organize, staff, lead, and control so as to reach them. Often, these actions are called the function of management. Sometimes, management is defined as reaching goals through people. This is a good definition because it underscores that managers alone cannot make an operation a success; it takes others working with management to accomplish this. Some workers see management as the boss giving the orders, as the one getting the most pay for doing the least work, or as the sector of the company that knows less about the operation than anyone else.

Such opinions are often based on a lack of knowledge about what the manager is or does. Management's failure to inform members of the operation about its essential role can be costly to its image. Good management often consists of the ability to plan operations intelligently and then to select and direct people so that the plans come to a successful conclusion. The more effectively managers prevail upon others to do a good job, the better management succeeds in its job. Many people are involved in achieving an enterprise's goals, and management is responsible for getting people's energy going in the right direction.

The hierarchy in a foodservice has three levels: management; mid-management or supervisory; and workers. To maintain a smooth-functioning facility, employees at each level must adequately perform the essential functions allocated to that level. Management comprises the chairman and members of the board in a corporation, the president, the manager, and even the assistant managers. Midmanagement may reach as high in the organization as the assistant managers and may extend as low as supervisors; sales managers, food and beverage managers, and purchasing department heads are typically considered mid-management, as are chefs, chef stewards, the maître d'hôtel, the catering manager, a food production manager, and a dietitian in charge of a hospital's dietary department. The workers are everybody else.

In a traditional continental-cuisine organization, the executive chef is a mid-management person under a food and beverage manager or other manager. The sous chef is a supervisor under the executive chef, and the chefs des partis are sub-supervisors under the sous chef. Under these people come workers such as cooks, helpers, and others working in the organization. In most foodservices, the number of people at the worker level far exceeds the number of people at the other two levels.

The management level plans action, the midmanagement initiates and supervises action, and the workers carry it out. It is important that all employees in the organization know what type of action they are responsible for. When people at a lower level try to assume the functions of others at a higher level (or vice versa), conflict can occur. The intruding person is resented as encroaching on someone else's turf. A person promoted from one level to a higher one sometimes encounters problems in this regard because he or she finds it difficult to break away from the action patterns learned in the lower level. This is seen when a supervisor promoted from worker ranks fails to supervise without constantly pitching in and doing jobs others should be doing.

Job Opportunities

The employment outlook in foodservice is excellent. The

reason for this is due to the unprecedented growth in the industry over the past two decades, resulting in exceptional opportunities, especially for the graduates of management programs. The managerial job pool is growing at a rate better than the rest of the economy. In some segments, such as contract foodservice and quick-service restaurants, the need for entry-level management is acute. New opportunities will arise in management for those willing to serve the "baby boomers" as they age in assisted-care living facilities.

With the trend of foodservices being increasingly dominated by large national and multinational corporations, the demand for managers with a formal education will continue to increase. At the same time, the industry will provide a vast array of opportunities for those people who possess a strong entrepreneurial spirit. For innovative, hard-working, service-oriented people who want to own their own business, foodservice offers almost unlimited potential. Graduates of management programs typically begin their careers as management trainees, assistant managers, or supervisors. Starting salaries are competitive with those in other industries, but the potential for advancement significantly exceeds those in other industries for the capable person who is willing to work hard. Opportunities also exist for graduates in the areas of accounting, marketing and sales, finance, and human resource management. Of course, foodservice consulting as a career for the seasoned industry professional can be very rewarding.

In deciding if a career in foodservice is for them, individuals must be people- and service-oriented. Hours to be worked are often during hours when others are relaxing or at play. Relocating for the right job may take a foodservice manager across the country. Finally, to ensure a smooth-running operation, the foodservice manager often works more than a 40-hour week. Despite the hard work, the rewards can be great for those seeking opportunities for professional growth, good salaries, responsibility, and high levels of personal satisfaction.

Getting people into the right place so they do the right job is called "setting up an organization." A business organization may be defined as an arrangement of people in jobs designed to accomplish the goals of the operation. Organizational structures vary according to the needs of the enterprise, the jobs done, and the people in the organization. Organizations should be dynamic, not static. They must be changed as needed, and organizational charts should reflect such changes. Obviously, just setting up an organizational chart does not build an organization. Creating the living organization-by selecting and organizing people on the job-is one of the most important responsibilities of management.

Types of Organizations

Different organizations are needed to do different things for operations. Basically, there are three kinds of organizations: line, staff, and functional. Most organizations are combinations of these. For example, even though a line organization may be set up, parts of the operation may be strictly based on functional tasks or on staff work that is required to support the line operation.

Line organization

Military organizations usually are considered the best examples of line organizations. The lines of authority and responsibility flow from the head down through the various mid-management layers to the workers. In a line organization, a manager directs an assistant manager to do something; that person in turn directs a supervisor, who in turn directs a worker or workers. In reverse order, the report of work done or problems encountered flows back up through the same line of authority until it reaches the manager. No bypassing of layers is permitted either on the way down or on the way up. Sometimes this flow of authority and responsibility is called the chain of command.

In a line organization, the top position bears full responsibility for everything that happens in the organization. Such responsibility cannot be delegated to others. The power to make a decision or do a job can be vested in a subordinate, but the way this duty is discharged by the subordinate remains the responsibility of the chief officer. A good rule is "you can delegate power but you can't delegate responsibility."

Staff organization

Staff organizations are usually support organizations established to render a service to the line organization. The staff does not take steps to see that things are done, nor does it have authority to implement any decisions it reaches or to act on any research it gathers. These are tasks for the line organization to perform. Every organization has staff positions-such as a secretary who works for the manager-that carry no line authority. It is improper for such a person to give orders to others in the line. Another staff organization is the accounting department, which renders a service to management by keeping records and furnishing essential financial and accounting data. A consulting dietitian serves in a staff position in relation to the superintendent of a small hospital or nursing home. Much trouble can occur when staff members step out of their staff roles and assume an unwarranted line role and give orders.

Functional organization

If a large corporation were to set up its organizational structure on a functional basis, it might set up a division that was in charge of sales, promotion, and advertising; another that was in charge of planning; another that was in charge of procurement; and so on-seeing that all functions needed to bring the operation to its goals were established. The various functional divisions of a corporation coordinate their activities and (usually) report to a high-level line organization. It is possible for line organizations to exist under a functional one. Thus, individual operating foodservices that receive instructions from various functional divisions are set up as line organizations. Functional organizations are best suited to large corporate units.

Management Skills

In any foodservice, daily events raise the need for action by management. Management must be prepared to see that the proper decisions are made so that the operation functions efficiently. Poorly devised action or inaction by management can ultimately cause an operation to fail. Proper procedures and records must be maintained. Employees must be selected with care and given adequate training. Personnel records must be established and maintained, complete with job descriptions. Supervisors must be properly instructed as to plans and actions and must be given adequate authority and backing to ensure that the work progresses satisfactorily. Planning, production, and service must flow smoothly so that patrons are satisfied and functions occur when they should. Making the operation run like clockwork requires considerable management skill. Sound decision making is the product of good judgment, practical knowledge, and astute thinking. The foodservice business is a dynamic, fast-changing one, and management must be equal to its demands.

Managemet by Computer

The job of managing has been considerably assisted by the computer, which has changed many clerical tasks from manual to mechanical processes, giving management more accurate and more reliable information in a shorter time. This permits better decision making and faster action in response to changing conditions or unexpected developments. The overall result of the computer revolution has been better planning, improved organizational control, and more up-to-date and more nearly complete information. Many tasks are simplified, and some are performed almost automatically. Consequently, management has more time to devote to planning and to direct supervision.

Although major computer application came relatively late to the foodservice industry, it is now being extended rapidly into many functions. Although various computer applications are presented, few (if any) operations have yet put together computer programs that include all of the things a computer can do that are discussed here. The use of computers and the coverage they provide in foodservice operations will undoubtedly continue to be extended in the future.

Menu planning

Early use of the computer in foodservice operations came in the areas of payroll, inventory, accounting, and point-of-sale transactions-largely because systems covering these areas could easily be adapted from other industry computer applications. The first original programs unique to the foodservice industry were programs set up in the 1950s and 1960s for computer planning of menus. These early programs had as their goal the production of menus planned to meet nutritional standards within certain cost constraints; they were largely developed for health-related foodservices.

Today, menu-planning programs are used to produce menus that meet specific patron and operational needs within certain cost limitations. The program can be set to offer a desirable range of individual menu items, such as the following four soups: one a cream soup; another clear; another heavy with vegetables and other items; and the fourth a purée. The menu programming can include instructions to avoid too much repetition of ingredients or flavors, such as having tomatoes appear a limited number of times in dishes on the menu. Different forms of preparation can also be specified, to ensure that different cooking methods are found in the new items offered, such as a braised meat item, a roast, a broiled meat, and a deep-fried meat item.

The computer can also be programmed to present a balance in meat, fish, and poultry offerings. Special combinations can be established, such as having Yorkshire pudding always appear automatically whenever roast ribs of beef are on the menu, having a horseradish sauce always appear with boiled tongue, or having deep-fried onion rings always appear with sautéed calf's liver. The frequency with which menu items reappear from week to week can also be set. If a retirement home wants chicken to appear on the menu for every Sunday dinner, the computer can be instructed to make this happen. The computer can give a printout of nutritional values for meals so that management can monitor them to ensure that the proper nutrients are being served.

Recipe size can be expanded or reduced automatically. Thus, a recipe can be exploded (increased) from 25 portions to 150, 250, or 1000 just by giving the computer the right instructions. Other constraints or needs can be input to the computer and it will comply with them in generating its menus. Quite flexible programs are possible, and often management can override the computer by adding, deleting, or otherwise changing what it does. The list of menu items that is entered into the computer to use in menu planning is called the menu file. Normally, each menu item is identified in the computer by its code number, but some computers can identify items by name as well.

Many menu items in the menu file require special preparation before service, and so must be backed up by one or more recipes. The recipes listed in this manner are kept on what is called a recipe file. Thus, an item titled Baked Cod Mornay may have two recipes: one for baking the cod in a mornay sauce; and one for making the mornay sauce, which is a béchamel (cream) sauce to which cheese has been added. Thus the mornay sauce in the Baked Cod Mornay recipe actually becomes an ingredient in the baked cod recipe. These other recipes for sauces and other items supporting the main recipe are sometimes called subrecipes. Recipes should be standardized to give a known quantity and a known quality at a known cost. Stan-dardization is accomplished by pretesting recipes and precalculating costs.

Recipes should specify the code number or code name, the total amount of food produced and/or the number of portions, the portion size, the ingredients, the amount of each ingredient used, and the method of handling ingredients to produce the recipe. Cooking times, cooking temperatures, panning instructions, and other relevant information should also be given. Variations and substitutions can be listed at the bottom of the recipe. Total cost of the recipe and cost per portion are usually listed, and sometimes menu selling price is, too. In some cases it may be desirable to identify other items that are served with the item produced by that recipe or to include plating or serving instructions. A picture of how the item should appear is sometimes added, and some computers possess the necessary graphics capabilities for this.

Ingredient files

An ingredient file consists of all ingredients listed for recipes. Ingredients are identified in the computer either by their code numbers or their code names. The ingredient code number should either be the same as that used in purchasing or be closely related to it. Thus, a 6/10 case of tomato catsup might have a purchasing and receiving number of P2148, but on the inventory and issue records it is identified as 12148 because the shipping unit (the case) is opened and broken down in the storeroom and issued in individual No. 10 cans. Pricing also differs. The price of P2148 (a case composed of six cans) is $24.60, and the price of 12148 (a single can) is $4.10. The ingredient price and code may require a third set of numbers, since the catsup used in recipes is often measured by volume. Thus, in the recipe, the ingredient code may be R2148 and the cost $0.035 per fluid ounce.

If a recipe called for 11/2 pints of catsup, the computer would convert 11/2 pints into 24 fluid ounces and multiply 24 × $0.035, for a total ingredient cost of $0.84. Some computers cannot easily convert pints into fluid ounces; in such situations the recipe must give the percent of the can to be used. Similarly, in bar work, making conversions from liter bottles into ounce serving portions is difficult for some computers to do, in which case the bar recipes must express quantities of ingredients in milliliter portions. Changes in recipes or ingredient costs must be input as they occur so that the recipes can be updated and so that the information compiled by the computer reflects what is actually the case. Often a change in the purchase price of one ingredient automatically changes prices in other files. Recipes often specify the condition in which an ingredient is to be added to others during preparation of the item. For example, a recipe may call for the ingredient to be sifted, beaten, blanched, melted, diced, or strained.

Forecasting

To obtain an estimate of costs and/or profitability of a specific menu mix (group of menu items), management must develop a projection of the number of menu items that will be sold. This projection is called a forecast. Forecasting is still a rather inexact science. The many variables involved in future demand are often difficult to detect and evaluate. Historical data-such as amount sold for the same meal on the same day last week, last month, or last year-can be used. A popularity index can be kept to record how many portions were sold each time the item appeared on the menu. Another technique is to evaluate the other items competing against a particular menu item and estimate how much they add or detract from its sales. An item that might sell well when grouped with some items might not do as well against others.

The effect of weather, events such as conventions, athletic contests, and other independent factors may have to be taken into account. Food-eating trends and other long-term factors deserve consideration, as well. In some foodservices, fairly good estimates of sales can be developed by noting reservations or bookings (such as with hotels or airlines), occupancy projections for dormitories over weekends, normal attendance for sporting events, or other information that indicates what quantity of patrons to expect. Hospitals find that their patient count usually goes down before and during a holiday season, but hospitals have an advantage in forecasting that many do not, in that many have patients mark menus a day ahead, indicating the foods they want served to them.

Forecasting, if done with some care, is a better guide to future demand than just plain guessing or playing hunches, although these should not be ruled out entirely. With some practice and experience, prognosticators will find that their forecasts become increasingly precise and increasingly accurate. Once the forecast is made and input into the computer, the computer can generate cost, profit, and other financial data. The computer will read the menu item, select its proper recipe, calculate from the recipe the ingredient cost and the portion food cost, and multiply the latter cost by the number of portions given in the sales forecast to establish a projected material cost for the menu item. Repeating this process for each item on the menu yields a total food cost figure for the period covered in the projection.

If a selling price for each menu item is input into the computer, the forecast number of sales times the selling price gives the expected income from the item; again, when this is done for all menu items, the sum of the income figures will equal the total sales. These computations enable management to estimate how well the menu is going to do. If other cost projections are programmed, the computer can generate a net profit estimate and make other projections that are of great help to management.

Production Planning

An important part of managing a foodservice is production planning. There is a need to know when to produce foods, how much, when, where it will be needed, and other information. Unless such information is given, successful operation cannot be achieved. Such planning is the responsibility of top management, but it is often delegated to others.

The first step in planning production is to decide what is to be produced and when. This requires that a menu be planned and the recipe to be used be selected along with the date, meal, and sometimes, time in hours and minutes when it will be needed. The cost must also be determined to see if it fits into budget or menu price restraints. Sometime in this preliminary planning one must decide if adequate labor, equipment, and other facilities are available to produce what is needed. Also, is the amount of labor sufficient and sufficiently skilled to do the job? After all this, the decision of how much must come, and this requires a forecast. Next, a check must be made to see if the materials for production are on hand, and, if not, are ordered in the proper quality and quantity adequate to serve production needs. Proper receiving and storage must be given and the requisitions for supplies withdrawn from inventory must be made and delivered.

Production must occur at the proper time. This is frequently an important consideration in planning. Menu items take various times to be ready for service. An aspic must be made at least 4 hours before it is needed, to be set sufficiently for service. A tenderloin steak may be done to order. Bread and many bakery items are prepared hours ahead. Often, production requirements are such that the equipment used for production is pressed to produce what is needed and those doing the production must make sufficient adjustment to accomplish what is desired.

Sometimes, items can be prepared in advance, such as soups and sauces, and placed into a bain marie or steam table to await production. Planning is such that all units of a production department must at the time of a meal come together in unison with production complete. The start of a meal is like that of the start of a symphony orchestra in which every player is ready when the maestro gives the first stroke of the baton. It must proceed as an orchestra working in unity together. Although service is usually not a part of production planning, it must be considered. Successful production can succeed or fail with proper or improper service.

Proper dishup is needed. Portions must be sized properly and displayed attractively. The size of the dish holding the order or orders must be right, not too large nor too small. The portion should not come over the rim. It is said, "The rim belongs to management, and the rest to the server." Attractive dishware must be used. Dishware color and decoration should fit the environment of the facility. Thus, Chinese food should be served in Chinese-type dishware. Tableware should fit into the decor of the operation. The proper ware should be used.

If an upper-scale restaurant wants to feature Snails Bourguignon, they should be served on an indented snail plate with the right holding tongs and extractor. Otherwise, the dish loses much interest and may result in guest dissatisfaction. Food should be properly garnished. Although it is not possible for all garnishes to be edible, some might not be-paper frills used to cover the rib ends of lamb chops or of a pork crown roast, for instance. The garnish should not overwhelm the food but should serve as an attractive accompaniment to it.

Contrasting complementary colors between the garnish and food should be sought. The garnish should be fresh appearing, not "tired." Certain garnishes are typical of certain foods, and a garnish common to one type of ethnic food must be used with care with other types of goods. Unusualness should be sought. Thus, a bread or cream horseradish sauce is often an accompaniment to roast beef, but an unusual and interesting difference is to present the roast beef with several horseradish curls on the plate, omitting the sauce.

Types of Production Planning

Three methods of planning production are commonly used by foodservices, although there are others. These are the laissez-faire method, the production sheet method, and the ingredient room method. The first is common in many restaurants and other commercial foodservices. The second is common in institutional feeding and also in many commercial operations, while the third is common in institutions and in large-scale production units such as central commissaries. The laissez-faire method basically leaves most of production planning to those who will be doing it. Management or sub-management may indicate roughly some of the basic guidelines, such as menu items to be produced and time, but leave the remainder to employees. In many operations this works well.

However, in some, management may not even plan the menu. In a hotel, management usually leaves production planning to a food and beverage director or even to a chef. In a hospital the director seldom takes any action in production planning. This is left to dietitians or foodservice directors, or even at times those lower down in the organization. Where those actually doing the production also have the responsibility for production planning delegated to them, the lead employees with the responsibility usually have been employed in the operation for some time and know the needs or have sufficient experience in other foodservices to know what is needed if they are a new employee. Often, the person doing the purchasing confers with those doing the production to ascertain items needed, quality, amounts, and so on. They know from experience, for example, the times when planning must be done so that foods are on hand when needed.

Besides setting up for direct deliveries to supply the necessary items, some system of inventory withdrawal will be worked out either by management or the accounting division. Times for starting production are known from experience, and allocation of the work to be done is almost automatic. At times, those requiring items from another section may remind those in the section of this need, such as notifying the bakery section that so many vol-au-vent shells are needed for Tuesday's lunch. For this type of production planning to work, it is necessary that the entire operation work together as a team. The production sheet method usually requires that management or submanagement plan production, and usually the latter procedure is used. This requires that previous to production the menu and other steps be planned and then a sheet is drawn up detailing the production need, the amount to be produced, the time needed, and if food is delivered to different service areas, where such food is to go, when, and how much.

In a health facility, special diet needs may be noted, and production needs grouped under these categories.

Often, in a hospital general menu is planned, and from this the special diets are taken with the required modifications being made. The use of a special diet kitchen to prepare some diets is used today for only very special diets, and many health care operations have dispensed with them entirely. Usually, the production sheet also informs purchasing how much of specific items will be needed, although in some cases, this may be left to those producing the foods. Variations in this method are possible where experienced production employees are on the staff and certain decisions such as amounts to order, and so on, are left to them.

The ingredient room method requires that a file of standardized recipes be available, usually computerized. Mid-management staff usually have delegated to them the planning of the menu along with other preliminary planning decisions. This person often decides on the menu and the amounts to be produced. Once these decisions are made, the items needed along with amounts are fed into the computer and a printout of the quantities and quality of foods is obtained to send to the storeroom if needed from inventory, to purchasing if they must be ordered, to the vegetable preparation section, or to other basic supply units. These needs sent to these areas are divided by section or production area along with the time required.

The storeroom usually has a separate area or room where it weighs out or otherwise measures the amounts required for each recipe. This must be done with extreme accuracy in exact quantities to produce the desired product in the quantity required. Recipe amounts are then segregated by section following which the storeroom delivers the items to the areas where needed. Preparers then process the items according to the production sheet definitions. One requirement that must be impressed on preparers and enforced is that all ingredients delivered must be remeasured. This is to prevent the preparer from blaming the storeroom for failing to deliver the correct amount in the event of a recipe failure.

The ingredient room method has many advantages. It saves worker time. If a worker has to leave the station in which the person works and travel to the storeroom to get items, a high-priced worker is often spending time just walking and waiting to get the items. If a lower-paid storeroom clerk can do the job and deliver the items, the worker then stays in the work area and produces. It also gives good organization to production. It is obvious that such a system requires proper time planning. Menus and preliminary planning must be prepared sufficiently in advance for storerooms, vegetable preparation, or purchasing to have the items delivered when needed. However, once the system is worked out and put into motion, production usually proceeds smoothly, without a hitch.

Quality Control Points

A consistent level in the quality of menu items is achieved by establishing a standard for the product and then setting up checks at critical control points to see that the desired quality is achieved. These points should include correct ingredients and their measurement; preparation techniques; times and temperatures; correct tools, equipment, and utensils; condition of product during various stages of preparation; sanitation; portion size; and other factors required in special products. Each product will have its own standard, so there will be as many standards as there are menu items produced. It is important that the checks and procedures established should be in as much detail as the standard. Thus, recipes should detail the critical quality control points. Again, pictures of the product showing the right procedure or the results of the wrong one are helpful. If rather inexperienced personnel are to prepare a product, it might be well to go over the critical points before starting preparation.

Standard Formulation

A standard is something established as a rule or basis of comparison in measuring quality, quantity, value, or other factor of a product. In foods it covers factors such as color, volume, shape, crumb, crust, density or thickness, tenderness, juiciness, and so on-in fact, any attribute that affects the quality of a product. As noted, each product will have its own standards. Often, standards must differ according to the type of operation. Also, standards may even have to differ within a specific operation; thus, portion sizes may be varied between adults and small children. The number of standards required will depend on the number of items served in a facility. A drive-in selling only carbonated beverages, milk shakes and malts, hamburgers, and french fries will have only a few, whereas an upscale restaurant that changes menus for every meal, every day, and season to season will have many. Often, assistance in setting up standards for foods can be obtained.

Recipes often list many factors that one needs to cover in a standard, such as type, quality, and amount of ingredient; techniques to use; times and temperatures; utensils, equipment, and tools to use; condition of product during preparation, cooking, or baking; and as a finished product. Anyone with a good foundation in food preparation can thus take a recipe and formulate much of a standard. Often, food preparation texts give standards for products, and these can be used as reminders of factors to put into one's own standard for the product. Trade associations, purveyors of food products, and others often have standards for their products.

Advertisements of foods also give pictures of products and product descriptions that can be helpful. The federal government has many quality standards for foods. Thus, one can say in a standard for roast beef: "The quality and type of product should be a Range C, No. 109, upper Choice grade roast, aged in Cryovac 21 days." One then knows that it will be a beef roast from 19 to 22 pounds in weight, cut and tied ready to be roasted, of a specific quality to be satisfactory in flavor, juiciness, and tenderness to give good eating-these three factors being decided largely by the grade and aging specified-and sized to give a generous portion. Also, the federal government has many grade standards for foods that can be helpful in establishing factors to cover. Thus, a U.S. Department of Agriculture (USDA) score sheet for judging the quality of canned asparagus lists color, defects, character, and condition of the canning liquor as factors to judge. Using this, one would then proceed to add other critical quality factors required. In some of these federal standards for foods, one will find helpful information such as the syrup density or density of a product such as tomato puree.

In many operations personnel have the knowledge and skills necessary to produce the type of product desired. Seeking their help can often result in a good standard. One can stand and observe such a person preparing an item, noting the actions taken, and from time to time asking a question to explain why something must be done in a certain way. In judging products to decide on the standard to use, a panel of judges can be assembled and products scored. This is often a technique used to decide on the quality of liquors to serve as well as products. The panel should know the standard desired and what will be suitable to satisfy the patrons served. This brings us to the final determinant in what the standard should be-the patron.

A standard that satisfies the trade is essential, and no matter what management or anyone else in the operation believes, standards that do not satisfy will not keep a facility in business long. Thus, getting patrons to indicate what they like and do not like or might like or might not like can be extremely helpful and necessary in establishing a standard, and one must start from this premise. Otherwise, formulation of the standard may be a futile exercise.

Analysis of Menu

Menus can be analyzed in various ways, ranging from highly qualitative methods to complex mathematical ones. Subjective menu evaluation, menu counts (popularity indexes), Hurst's menu scoring, menu factor analysis, the matrix analyses of Miller, Smith-Kasavana, Pavesic, and Kotschevar, goal analysis, and break-even analysis are ten common approaches. Some involve radically different approaches and give different information; and even those that use similar methods vary in the information they give.

Customer Ordering

One of the more recent developments in computer use within quantity food production has been seen in customer ordering. While the fast-food and quickservice segments have led the way, other dining establishments are not far behind in their implementation of this latest hardware and software. Touch-sensitive screens are now used in operations where service counter personnel once took the order. Customers are moved through a series of graphics in which they literally touch the screen to select the food item desired. In the case of play areas where a mother wishes to watch her children without having to leave them to order and pick up food, touch-sensitive screens offer a method by which the mother identifies her table number and the food is brought to her.

Merchandising is enhanced through a voice that encourages the customer to order the largest drink since it is the best value. Since a tally is kept on the screen as to how much the customer has spent in the ordering process, customers are more likely to spend more in that they know how much money they have and how much they have spent in ordering. Thus, an extra apple pie or order of french fries are or are not purchased.

Devices are now also available that allow a server at the table to take orders, punch them into the device, and have the orders filled by the kitchen. When the order is ready for pickup, the device beeps; thus, the server is spared taking trips to the kitchen to place orders and to see if they are ready for pickup.

Just as the delivery of food to the mother watching her children in a restaurant's play area has been made possible through computers, so is the payment for food ordered at a touch-sensitive screen. By having storedvalue cards similar to those used to prepay a phone call through a calling card, the customer is able to pay for his or her meal electronically via computer. The system then allows the customer to order food without assistance, pay for it without interaction with another person, and pick up the food or have it delivered, as is the case in the play area. Ordering at drive-up windows is being revolutionized, as the "loud speaker order taker," who never seems to get the order right, is being replaced by a touch-sensitive screen that merchandises products through graphics, keeps a tally of the total dollar spent, and ensures that the correct items are ordered. Another drive-up window system has a "scoreboard" installed which lights up each item the customer has ordered. This innovation is linked to the point-of-sale personal computer to help ensure against communication glitches between the customer and order taker.

Computers will continue to affect the heart of the house delivery and payment systems as well. Purchasing agents will order on-line through the Internet. The future of the computer on food production, planning, and management will greatly affect the efficiency of operations, as has been seen in recent years.

The quantity of items to purchase or to issue is determinable when the quantities of ingredients needed to produce the menu are calculated by the computer. These quantities can be further distinguished as quantities needed for immediate use or as quantities to have on inventory. Timely compilation of such information can help operations avoid stock shortfalls and other frustrating food production problems. When an ordered item is received, the computer can be informed of the date, the item's code number, purveyor, amount, cost, a description of the goods, whether the item went directly to production (directs) or to stores (inventory), and so on. If programmed to do so, the computer adds the new amount to inventory and calculates a new inventory amount. Directs are charged out as, in effect, issues on the receiving date. By having the computer add directs to storeroom issues, management can obtain a food cost for the day; put into ratio with sales, this gives the percentage food cost.

Deducting issue values from inventory and adding receipts enables the computer to generate an updated inventory called a perpetual inventory. All these programmed tasks are rapidly and accurately performed by the computer, eliminating much slow and tedious manual manipulation and calculation. Much useful information for making management decisions can be obtained at the cash register, often called point of sale (POS). With a POS system, management at any time can ascertain what checks are out, how long checks have been out, what orders remain unfilled, how many items of a particular kind have been sold, what their dollar value is, what volume of sales has been generated by various servers, what amounts credit card charges, room charges, and tips come to, and many other things.

At the end of a recording period, total sales, total item or food group sales, and other desirable data are at hand. It is now possible for the computer to run an analysis of how well planning projections made prior to production anticipated actual results. This alone is of tremendous value to management as a tool for refining the processes of planning and decision making. In the days when manual compilation of such data was necessary, the results often were unavailable until too long after an actual occurrence for management to take corrective action. In addition, after all results are in, the computer can transfer essential data to the general ledger file so that important financial reports such as cash flow, balance sheet, and profit-and-loss report can be compiled. Thus, almost from cradle to grave, the computer can be effective in food production management. It has changed clerical, decision-making, and other timeconsuming tasks into almost automatic operations and has extended management's control and base of information so that better management is achieved.

Total Quality Management

Although the term total quality management has been used extensively, there has been some disagreement as to the nomenclature in use. Some refer to the subject matter as just quality management; others refer to it as continuous quality improvement. There is one aspect that has generated very little disagreement and that is the fact that management should focus on eliminating "quality" problems before they occur rather than attempting to address them after they have happened. This is especially true in the area of customer service but is also applicable in the area of food production. Both are important and must be given careful consideration.

The key to quality in service is to have a systematic, planned process for returning dissatisfied customers to a state of satisfaction after service has failed to live up to expectations. Since customer turnover can have a tremendous financial impact on an operation's bottom line, customer satisfaction is a must for total quality management. For example, the loss of a loyal customer is one thing, but the alienation of potential customers due to negative word-of-mouth referrals can be even worse.

Dissatisfied customers are likely to complain to nine or ten others about service imperfections, while satisfied customers typically express contentment to only four or five others. Furthermore, research indicates that it costs five times more to replace a customer than to retain one. Hospitals sometimes find that doctors and patients decide on which hospital to patronize because of high satisfaction with the food at that hospital. The management issue, then, is to empower service providers to engage in techniques to ensure customer satisfaction when, in fact, the customer is not satisfied. The fiscal and informational resources must be at hand for the server to see that the customer departs the operation satisfied.

Recovery actions often involve "doing it right the second time" but, more specifically, correcting what was done wrong immediately, so that when there is a second time, the customer notices that matters have been corrected. Recovery actions may take the form of compensation, apology, reparation, and empathy. Each must be understood by servers and applied correctly depending on the situation. Total quality management in the area of customer service is ongoing and must be given top priority.

The same holds true in food production. When the product produced or served is not up to the standard, the professional must recognize that fact and take corrective action. Checkpoints come from the cook, expediter, server, and customer. If at any point, one of those people says that the item produced is not what it should be, quality is affected and change must take place. Satisfaction can be improved by replacing the item immediately or offering the customer a different choice. When it comes to food production, the key is to correct the system such that another poor item is not served in the future. This may range from retraining the cook to regarnishing the plate. If it is the server's fault, such as serving cold food due to delay in pickup or taking down and serving the wrong order, the standards in the service area must be rechecked. Total quality management must be constant in any operation in order to raise the standard. Both customer service and food production are affected. Management must be aware of what is being done incorrectly and make every effort to improve in those areas.

Guidelines for Volunteers in Food Service

If possible, a person knowledgeable about food preparation and service should be designated to coordinate and supervise the food service activity. When individual contributions are solicited, food service activities should be planned and coordinated so that as much of the food as possible is prepared in one place.

According to Iowa law: churches, fraternal societies, or civic organizations which serve food one day per week do not need to be licensed. If these organizations serve more often than one day per calendar week at a single site, that site must be licensed. Schools and school-sponsored organizations are exempt from the license requirement as long as food is served in the school or at a school-sponsored event.

Dangerous Foods

Potentially hazardous foods are those containing milk or milk products, eggs, fish, meat, or poultry. They have the highest potential for causing foodborne illness because they can support rapid and progressive growth of harmful bacteria. Keep these foods above 140°F or below 40°F.

— Plan food service so that potentially hazardous foods are served within three hours of preparation.

— Use an accurate thermometer to maintain safe temperatures.

— Keep meats, gravies, and casseroles at an internal temperature of 165°F for serving. Keep other hot foods, such as vegetables, at an internal temperature above 140°F.

— Keep meat or poultry dressings (stuffing) below 40°F or above 140°F. Avoid the "danger zone" between 40° and 140°F. Never dress or stuff poultry the day before cooking and never partially cook a stuffed bird. Whole poultry must never be pre-cooked and then reheated prior to serving. Poultry, poultry stuffings, and stuffed meats must be cooked without interruption to heat all parts of the food to at least 165°F.

— Food to be served cold should be rapidly cooled to an internal temperature of 45°F or below. When practical, ingredients for cold foods should be chilled prior to use.

— Never allow raw poultry or meat to come into contact with other foods, especially those that will be eaten raw or are only slightly cooked.

— Leftovers and foods that are pre-cooked for later use, such as browned ground beef, should be cooled quickly so the center of the food reaches 45°F within four hours. This is impossible to achieve when food is placed in large containers. Instead, place the hot food in shallow, covered containers in the refrigerator.

— Egg-based products such as eggnog, cream puffs, cream pies, and puddings are potentially hazardous and should not be part of the menu unless they can be refrigerated immediately after preparation and kept refrigerated until served.

— Acid beverages, such as punch and fruit juice, must be stored and served in containers made from food-grade plastic, stainless steel, or glass. Never use plated, porcelain, or enamelware containers for storing or serving these beverages.

— Never lick your fingers or use the cooking spoon to taste while preparing food. Instead, use the "two spoon technique"-dip the cooking spoon in food, then transfer food to a tasting spoon.

As with any food service, common sense and good food handling practices should prevail. Keep in mind the rule "hot foods hot-cold foods cold." This will be easier if food is served in shallow containers and replenished frequently. Keep cold food iced or refrigerated. Hot trays are recommended over candle-type warmers that generally do not produce sufficient heat to prevent bacterial growth. Consider possible safety problems before using canned fuel-type heaters. Keep hot food covered as much as possible. Use separate serving utensils for each dish.

Bake Sales

Bake sales are not required to have a permit or prior health approval. However, they do represent a health concern. Organizers and sponsors should follow good sanitary practices when planning, organizing, and holding a bake sale. Care and thought must be given to preparation, transportation, display, and serving. Typical products suggested for bake sales include breads, cakes, doughnuts, pastries, buns, rolls, cookies, bars, biscuits, and pies (except meat, custard, and cream pies). Potentially hazardous foods should not be sold at bake sales. These include cream puffs, cream-filled pastries, cream pies, custards, and puddings.

When conducting a bake sale, follow these recommendations:

— Food should never be solicited or accepted from any person affected by a communicable disease or who is a known carrier of such a disease.

— Remind contributors that foods should be tightly wrapped or sealed and transported in clean vehicles. Foods should not be transported along with family pets.

— Maintain a list of contributors.

Include name, address, telephone number, name or type of food donated, list of ingredients used, and date food was prepared.

5

Restaurant Business

In the restaurant business, you have so much fun that you can hardly stand it. You get tired of wrapping the presents, preparing the eggnog, and checking that the elves are on time for their shifts, and if you have to look at any more roasted chestnuts, you'll die. The restaurant business quickly becomes more work than fun, so don't be fooled. The restaurant world is more than glitz and glamour. It's truly a business, and if you don't look at it that way, you won't succeed. Ultimately, being a restaurateur is being a manufacturer. You're producing a product (food) from raw materials and selling it to a customer. You're competing with lots of other "manufacturers" for that same diner. So you better do it better than the other guy, or you'll be out of business.

Sometimes the business of the business is tough for people to relate to. It's a hard concept for many people to get because your product isn't packaged in a box that sits on a shelf. Your product is packaged in many layers—including your exterior, your lobby, your staff's attire, the music playing, the aromas emanating from the kitchen, the friendliness and knowledge of your staff, your silverware, your china, and your glassware. All these things make up your packaging, affect the costs of doing business, and affect your diner's decision to come in and, ultimately, to come back. As with any business, the planning stage is crucial, and you have to survive it before you can enjoy any of the fun.

Right off the bat, you have to create a timeline for getting your business up and running, develop your restaurant's theme and concept, research the market, develop a detailed business plan and use it to find and secure financing, and find the best location for your new restaurant and get the right licenses and permits. Buy your products at the right price and sell them at the right price. This simple tenet can make or break your business.

Depending on how new you are to the restaurant biz, you may need accountants, attorneys, contractors, and host of other characters, all at the ready and working with you at various stages of the project. Hire an accountant early in the process of setting up your business. She can help you get your numbers together for your business plan, which is a must do if you're trying to get financing for your venture. After you're up and running, you'll analyse your monthly financial reports and look for ways to improve the numbers. A good accountant, preferably one with restaurant experience, can help. When starting any new business, you'll need to review contracts, file your permits, or maybe incorporate your business.

Depending on how you set up your business, you may need to draft a partnership agreement or two. Before you sign franchise agreements or vendor contracts or fire your first employee, make sure that you're working with a good attorney, who can help you with all these tasks and more. Most people starting a new restaurant, or taking over an existing one, change a few things (or a few hundred things) at their new location. May be you need to set up a new kitchen from scratch or improve the air flow of the hood over the range. Maybe you want to upgrade the plumbing or install air filtration in your bar. Contractors can save you lots of time and trouble. Don't hesitate to ask them questions and check their references. Interior designers and architects come in very handy around renovation and revamp time. Sometimes they can come in and give your place a face-lift for much less than you might imagine.

All the hard work that's required to get to the point where you can open the doors will mean absolutely nothing if no one shows up. You have to start thinking about how to draw customers way before you open your doors. Develop your marketing plan based on what's special, unique, or different about your restaurant. Maybe it's the food, ambience, price, or value. Study your competition, watch what they're doing well, and understand where you have the advantage. Different groups respond to different messages. After you get the customers in the seats, you have to keep them there.

To be successful in this or in any business, you need to take care of your business today, tomorrow, and years from now. Stay up on trends in your sector and the restaurant business as a whole. Watch for information about shifting dining preferences and behaviour in trade magazines, print publications, television news (and the not-so-news magazine shows), the Internet, or anywhere else you get information. And always keep an eye on your competition. Don't copy them, but know what they're up to. It takes way more to run a restaurant successfully. And that's anyone can run a restaurant, but not everyone can run one well.

Success in Restaurant Business

This is a tough business, and if you want to succeed, you have to have the inner motivation - the drive - to sustain you through all the downs that accompany the ups. This isn't a venture for the faint of heart. If you want to own a restaurant to have a place to hang out with your friends and get free drinks, take the bar bill and avoid the hassles. The first thing you need to do, before you invest any additional time or money in this venture, is to examine and understand the factors that motivate you. Be honest with yourself.

There are lots of great reasons to want to run a restaurant. They are:

— You love an ever-changing work environment.

— You love taking on a challenge.

— You're passionate about the business.

— You have a passion for food.

— You hate having any free time (including the holidays).

— You're continuing the family tradition.

And the following list contains a few reasons that should send up a red flag in your mind:

— You think it will be fun.

— You want to be a celebrity chef.

— You want a place to hang out.

— If Emeil can do it, so can you.

— You're tired of having a "real" job.

— You've always wanted to run a restaurant after you retire.

If one or more of these reasons sounds familiar, don't be completely discouraged. Just make sure that motivations such as these aren't your only, or even your primary, reasons for wanting to get into the business. And do some further investigation before making the financial, personal, and professional commitment to the business.

Running a restaurant, either yours or someone else's, is a huge commitment. It requires long hours, constant vigilance, and the ability to control potentially chaotic situations—on a daily basis. Running a restaurant shouldn't be a leap of faith. You need to go into this with your eyes open. You carefully consider your motivations, you also need to make sure that your expectations are firmly planted in reality. Take out a pen and some paper. Divide the paper into two columns. In the first, list all of your expectations for the future business. From the profits you expect, to the lifestyle you hope those profits will support, to newspaper reviews or the customer views you hope to elicit, list it all. This is your chance to put your dreams on paper. Then, in the second column, write down what you expect out of yourself to make this thing happen—your contribution in terms of time and money, sacrifices you'll have to make, and anything else that you can think of.

Tracking Key Traits

Based on experience in the restaurant business, successful restaurateurs exhibit a few common traits.

Business sense

This is probably the single most important trait. For all that the restaurant business is, it's still basically a business, subject to the same pressures as any other. Keep that thought in mind going into your arrangement. If you don't, you'll be hard-pressed to succeed. Skills that you've learned, developed, and honed in the real world can apply in this business, like buying skilfully, managing tactfully, and negotiating shrewdly. But many different facets of this business are tough to pick up.

Tolerance

The ability to keep your cool under pressure, thrive in chaos, and handle multiple points of view and personalities will serve you well in the business. Whether you're dealing with customers, employees, purveyors, changing trends, or a fickle clientele, you have to develop a thick skin. The inherent stress of the restaurant makes for short fuses. Your job is to dampen those tempers, smooth the rocky waters, and calm the storm.

Flexibility

The environment changes from minute to minute. You have to be able to adjust and think on your feet. You have to have a good balance of process- and product-motivated people. Process-motivated people micromanage what's going on in their organization. Product-minded people focus on the end result. Sometimes you'll wear both hats.

Creativity

Infuse creativity into every facet of your business from how you approach your customers and your food, to how you promote your business. That creativity affects how your business performs.

Positive energy

W. ıever you're in the restaurant, you have to be "on" - all the time. Restaurants that have a positive vibe are the ones that make it. Positive energy is the differentiating factor, as intangible as it is, between the winners and the losers in this business. And your restaurant can't have it if you don't.

Ability to hold (or hold off) liquor

Coveted by many, achieved by few, the ability to handle one's liquor has been the downfall (physically, financially, and spiritually) of many a restaurateur. Per capita, no industry drinks more than the restaurant business. For some people, managing a restaurant is like getting the keys to the grown-up candy store, and the temptation is too much to resist. As a restaurateur, you often drink as part of your job. No matter what the circumstances, you still have to count the money at the end of the night, or you have to be ready to go first thing in the morning.

Leadership skills

Being a leader in this industry is being able to balance an entire range of different management approaches, knowing when to lead by example, and knowing when to give the troops their marching orders. Most importantly, a successful restaurant leader is able to find her own leadership style and deal with employees fairly, consistently, and with respect.

Schmoozability

People like to feel important. They want to be part of the inner circle of the restaurant, no matter how large that circle may be. It's cool to say, "I know the manager" or "The chef's a friend of mine." Nothing gets return business like calling a diner by name. That's why you put up with the pictures of grandkids, whines about their big project at work, or not-so interesting travel tales. Always make the customer feel welcome, at home, and at ease. Turning a good mood into a bad one is incredibly easy. Turning a bad mood into a good mood is exponentially more difficult.

Passion

To succeed, you have to have it. Running a restaurant is a business that eventually chooses you; you ultimately can't choose it. If you don't have passion for the business, you can't sustain, maintain, and overcome the obstacles that crop up. You can't teach it or learn it; you have to feel it. If you don't have the passion, your task of becoming successful will be exponentially harder. You have to connect everything to your passion. You have to get the wait staff wired with it, because they're selling your vision to the customers.

You have to get the prep guys pumped, because they're cranking on a tough schedule, without the natural excitement of a restaurant full of people. You have to get the dishwashers psyched about cleaning the dishes, because the dishes frame the experience for the customer. Diners should experience a buildup of expectation for their experience from the first time they come into contact with anyone from the restaurant. Imagine doing all that without a passion for your restaurant, and you see why passion is mandatory.

Presence

Being there day in and day out has no substitute. Absentee landlords need not apply. Just stopping in to say hello or giving off an aura that you know what's up ultimately won't allow you to run the restaurant. If you're not there, those who are there in your stead will be the de facto rulers, and it won't be the same. If you're not physically present in the building most of the time, the schmoozing, the energy, the passion, and so on can't get to your staff and ultimately to your diners. You can't positively impact your restaurant if you're not there.

Equipment by Functional Areas

Receiving is a functional area and the first step in the flow of food through a foodservice system. The receiving area functions as a control point for most food and a critical control point (CCP) for processed and ready-to-eat food. Regardless of the size of the foodservice system, the activities associated with receiving are the same. The purpose of these activities is to ensure food quality and food safety. The activities of the receiving function include:

— Visually inspect all items and look for signs of contamination or container damage.

— Check expiration and pack dates.

— Count boxes invoiced but not delivered; note shortages.

— Check the product number against the purchase order or receiving ticket. Only accept approved brands.

— Check substitutions for approved brands.

— Check temperatures and record on chart.

— Remove any loose staples or fasteners.

— Reject unacceptable goods and note on invoice.

— Sign the invoice and retain a copy.

The objective of good receiving practices is to certify that all food moved into production is an approved brand. In addition, receiving is the opportunity in the flow of food to determine that products are in good condition, appropriate temperature, and delivered as ordered. Suggested equipment for the receiving area in a conventional kitchen includes:

— 2 wheeled hand truck and/or

— 4 wheeled platform hand truck

— utility carts

— table/desk for paperwork

— scales

— thermometers

— fly fan

Dry Strorage and Chemical Storage

After receiving food and supplies the foodservice assistant moves the goods to dry, refrigerator, and freezer storage. This is the second functional area. Dry strorage and chemical storage, although similar in function, must have all contents stored separately. Food must never be mixed in a storage area with chemicals or cleaners.

The dry storage area should be clean, dry, well ventilated, and temperature controlled. The recommended dry storage temperature is between 50° F - 70° F. Shelving selection will impact air circulation, pest prevention, and first-in first out (FIFO). There are a variety of shelving options.

Style and finish of shelving should be consistent with storage application and budget. Style examples include:

— open wire

— louvered

— embossed or raised steel

— solid

— vented, steel reinforced plastic

— polymer composite

Finish examples include:

— galvanized - solid, embossed or louvered

— zinc

— chrome

— zinc plating and organic coating

Shelving Systems

Types of shelving systems:

— stationary - put in place and remain for a period of time, lacks versatility

— mobile - similar to stationary but has casters

— track - also known as high density or active aisle shelving, derivative of mobile system with mount or guided on a track (floor or ceiling)

— wall mounted - adjustable cantilever shelving - adjust to various heights, ease of cleaning

Suggested equipment for the dry storage area in a conventional kitchen includes:

— shelving system

— dunnage racks

— can storage rack (optional)

— utility carts

Food and chemicals are stored separately to prevent any possibility of chemical poisoning. Your local chemical supplier can make recommendations for chemical storage organization. Suggested equipment for the chemical storage area in a conventional kitchen includes:

— shelving system

— janitor sink

— washer/dryer (optional)

Cold Food Production

Cold food production is a functional area that requires refrigeration, equipment, water, and a work area. A variety of activities take place in this area from washing produce for preparation to slicing, chopping, mixing, and plating. The equipment described for cold food production in a conventional kitchen includes:

— refrigerator/freezer

— blast chiller

— ice machine

— mixer

— slicer

— cutter/mixer

Employee competence to operate a slicer, cutter/mixer, or other equipment should be certified before the equipment is used.

Most refrigeration and freezer systems have a compressor, a condenser (air or water cooled), an evaporator, and a fluid called refrigerant. This system moves heat out of food via moving air. The purpose of the refrigerator is:

— keep food cold
— reduce the rate of deterioration by diminishing enzyme activity
— control the growth of microorganisms that potentially can cause foodborne illness
— maintain food quality

Walk-in refrigerator

The walk-in refrigerator or freezer is designed for bulk storage. Virtually any size is available. Most walk-ins today are pre-engineered (prefabricated) panelized construction which makes them flexible within the space available. Constructed of a series of 4" (1.0 cm) thick manufactured and modular panels, individual panels are attached with a variety of latches, camlocks, or bolts to form outer walls, ceilings, and in some cases the floor. There is a variety of heights, lengths, and widths available. These are specified either self-contained, quickconnect, or remote refrigerator system. Sufficient air space must be available around cooling units to prevent build up of excess heat.

Reach-in Refrigerator:

— versatile
— reasonable cost
— storage efficiency

One, two, three compartment - many options available depending on:

— budget
— capacity
— general storage, special products
— should eliminate interior shelves to adapt refrigerator for roll-in carts

Determine capacity and size by:
— approximate number of meals
— receiving schedule
— kitchen space available

Blast chiller

The blast chiller is designed to rapidly cool almost any kind of food. Once thought to be only needed in a cook-chill system, the blast chiller has taken on a new place in food production. An essential component of a Hazard Analysis Critical Control Point (HACCP) system is time-temperature holding of food. This includes proper cooling. The data verifies that the primary cause of foodborne illness in the United States is improper cooling of hot food as bacteria thrive in the temperature danger zone (40° F - 140° F). There are recommended procedures to cool food. However, if use of one of them is not routinely used, a blast chiller is recommended.

Cooling hot food in a foodservice walk-in or reach-in refrigerator can pose a serious risk of foodborne illness. These units are designed to store food that is already cold. They were not designed to chill hot food. Refrigerators don't create enough air flow to pull hot food through the danger zone to below 40° F in the recommended amount of time. The capacity of the blast chiller (in pounds) per load equals the number of persons the chiller can serve for three meals per day.

Ice machine

Commercial ice machines are actually small manufacturing plants. They utilize water and electricity to make ice. Ice is considered a food and must be handled like a food.

Ice machines are rated according to the amount of ice the machine produces within 24 hour period. Consider, storage bin plus the amount the ice machine can produce during the peak period = total ice available. A storage bin must be selected that is slightly larger than the machine's rated production capacity (risk ice melt in bottom). Carefully select a bin to meet production and service needs. Three types of condensers:

- air cooled - generally cost less and easy to install, but tends to heat environment around them
- water cooled - uses water to remove heat, no fan, quiet to run
- remote air cooled - heat dispelled through a vent, usually located on the roof, more costly to install but can reduce monthly bills

Mixer

The mixer is considered a valuable and versatile piece of equipment because of the variety of attachments. There are numerous manufacturers and models of mixers on the market. Planetary action mixing provides thorough blending and mixing action by repeatedly turning the mixed product into the center of the bowl. Mixers range from table top to 140 quart models geared to high volume institutional use. Mixers have a wide application with standard and optional attachments to meet the needs of changing and expanding food service menus. Mixers come in a variety of sizes/capacities, ranging from 5 quart to 140 quart, indicating the bowl size. The applications determine the correct size needed.

Slicer

There are numerous manufacturers of slicers on the market. Most are designed to slice high volume amounts of meat and cheese quickly and efficiently. The slicer (manual or automatic) may be used for:

— slicing hot or cold meat in uniform thickness

— slicing cheese in uniform thickness

— slicing vegetables with optional accessories

Options and accessories available at extra cost:

— optional carriage fences to slice tomatoes

— food chutes to slice elongated vegetables

— food receiving tray

Cutter/Mixer

Also known as vertical cutter mixer (VCM), this equipment:

— prepares food quickly

— versatile piece of equipment; based on blender concept

— consists of conical mixing bowl with motor mounted at the bottom.

— removable sleeve with cutting blades mounted on the motor shaft which projects from the bottom of the bowl

Hot Food Production

Hot food production is a key area in the conventional foodservice system. Activities that take place within the cooking area are grouped according to the type of treatment required. For hot food that includes:

— seasoning

— mixing

— shaping

— breading

— panning

— cooking

The selection of type of equipment is affected by factors like:

— food to be processed (menu)

— type needed (bake, fry, steam)

— labor availability and skill

— serving schedule

Heat is applied to food to increase digestibility and customer appeal. Nutrient retention and food safety are always considerations. Heat is transferred to/from a product in four ways:

— conduction - molecule to molecule, example: pot or range top, oven, steam

— convection - heat transmitted through liquid or gas to food, example: convection oven

— radiation - radiant waves

— induction - raise heat because pan is subjected to alternating electromagnetic field; with this process the operator must use stainless steel cookware.

In the hot food production area, attention to internal cooking temperature, time-temperature monitoring, and minimum time in the danger zone are all important. The following equipment described for hot food production in a conventional kitchen includes:

— tilting braising pan

— steam jacketed kettle

— direct steam kettle

— convection steamer

— pressure/pressureless steamers

— pressure steamers

— high pressure steamers

— combination oven/steamer

— convection ovens

— fryers

— range

Tilting braising pan

The tilting braising pan is also known as tilt skillet or tilting frypan. It is a one piece stainless steel pan with vertical side walls, a front pouring lip, and hinged stainless steel cover; it is also considered a griddle with high sides and a cover. The tilting braising pan has a variety of applications like: braise, grill, pan fry, saute, steam, and boil Examples:

— brown or braise stew meats, ground beef

-- shallow pan fry chicken, fish

— saute vegetables, stir fry

— cook soups, stews, sauces

— back-up grills, steamers, kettles

— steam vegetables with optional steamer pan insert rack

Steam jacketed kettle

A steam jacketed kettle is cylindrical in shape with a rounded bottom. A second larger cylinder is placed around the outside of the main cylinder creating a space between the two. This space is sealed so that steam may be injected. As steam enters this space it releases its energy (heat) by condensing onto the inner kettle's outer wall. This transfers heat to the food being cooked. Consider a double boiler used in a home kitchen. A steam jacketed kettle is the same basic concept.

There are two designs in kettles - a fully jacketed and a two-thirds jacketed kettle. Within the basic design of the fully jacketed kettle, there are two varying designs in the interior construction. One design is described as a hemispherical bottom interior. The other is described as a dish type slopped interior. Kettles are available in direct steam or self-contained electric or gas that are tilting or stationary.

All stationary models are supplied with a food draw-off system. There are two designs and varying sizes available on the draw-off valve, the tapered plug type, and a screw type gate valve (compression valve). The steam jacketed kettle is used for boiling, stewing, simmering, braising, holding, and rethermalizing (reheating).

Direct steam kettle

Direct steam kettles operate with steam provided by a remote steam source. An optional piece of equipment is the mixer kettle. Agitator and scraper assemblies are available for most direct and selfcontained electric, table top, and floor mounted kettles. This kettle, also referred to as cooker/mixer, features a mechanical mixing unit which can eliminate the manual stirring of ingredients. This saves labor time. The mixing of the agitators/scrapers reduces cooking times.

Wall-mounted kettles can be mounted and installed on inwall support arms (wall carriers). This feature provides for optimum sanitation. This type of installation may be seen in conjunction with energy distribution systems.

Convection steamer

Each manufacturer varies in how steam is supplied to the cooking compartments and what controls are used to perform this function. Pressureless steamers are available with selfgenerating gas or electric steam coil generators and direct steam models. The steam is supplied from an outside source. Direct steam units require a clean, potable steam supply. Perforated pans are preferred when possible for more consistent cooking but solid pans are acceptable. Capacities vary among equipment manufacturers. These steamers require a water supply which fills the steam generator/boiler and supplies the cold water condensate system. Open air gap floor drains are also required within a specific distance from the steamer.

Pressure/Pressureless steamer

These steamers offer the flexibility to operate in a pressure steam mode or a pressureless steam mode. There are two types of pressure/pressureless steamers manufactured. One type offers two or three cooking compartments where the top compartment is converted from a pressure to a pressureless mode and the bottom cooking compartments always operate in a pressure mode. The pressure mode for this type operates at 5 or 6 psi depending on the manufacturer. The steam temperature is approximately 228° F (109° C). The other type offers two cooking compartments where both may be converted from a 10 psi pressure mode to a pressureless mode.

The steam temperature is 240° F (116° C). These steamers are generally mounted on a 36" wide cabinetized base. Steamers are available in self-generating electric and gas steam generator models. They are also available in steam coil generators and direct steam models whereby the steam is applied from an outside source. Direct steam units require a clean, potable steam supply. Self-generating pressure steamers may be specified to supply steam to an auxiliary piece of equipment such as a kettle. It is used for steam cooking, defrosting, and rethermalizing.

Pressure steamer

There are two categories of pressure steamers manufactured. One category is known as low pressure steamers and the second as high pressure steamers. Each will be described separately. Low pressure steamers are usually mounted on a 36" wide cabinet base and have two to three cooking compartments. Steamers are available in self-generating electric and gas steam generator models. They are also available in steam coil generators and direct steam models whereby the steam is supplied from an outside source. Direct steam units require a clean, potable steam supply. Direct steam units are also available with four cooking compartments.

Pressure steamers are recommended for cooking like foods per batch or compartment load, as flavors may transfer. Self-generating pressure steamers may be specified to supply steam to an auxiliary piece of equipment such as a kettle. High pressure steamers are offered in counter top, stand, or cabinet base mounted models. There are several model sizes manufactured. Steamers are available in self-contained electric, gas, or direct steam models. They are also available mounted on electric, gas, or steam coil generator (boiler) in a cabinet base.

The counter top models require no water or drain connection. Water is filled and drained manually. Generator (boiler) based models require a hot and cold water supply and open air gap floor drain connection. The operation is automatic. The self-contained countertop models require a heat-up time on each cooking cycle. The steam generator (boiler) based units require an initial heat-up time for the steam generator then steam is readily available for each cooking cycle. The cooking compartment door is an inside self-sealing type and cannot be opened during a cooking cycle.

Combination oven/Steamer

The combination oven/steamer offers a single cooking chamber with the ability to cook in three cooking modes which include hot air mode (convection oven), a steam mode, and a combination of both modes. The combination oven/steamer may cook in all three modes independently or in sequence moving from one mode to another depending on the menu item, i.e., starting the cooking cycle of a roast in the steam mode to sear the outside and accelerate the cooking process, move to the combination mode to reduce shrinkage and the cooking time, and finish in a high temperature convection hot air mode to enhance the outside color.

Combination oven/steamers are available in gas or electric models. They are available in two generic sizes which are half size and full size. It is used to roast, steam, bake, poach, blanch, grill, defrost, cook-and-hold, and rethermalize.

Convection oven

The convection oven offers a cooking chamber with a fan to circulate hot air. This is called forced convection heat which is the origin of the convection oven name. Convection ovens are available in gas and electric models. Convection ovens are offered in single, double stacked, and roll-in models. Each cooking chamber has 11 rack guides and is supplied with five racks. It is used to bake, roast, slow roast and hold, and rethermalize.

Fryer

There are three basic components to every fryer:

— the fry tank

— the burners heating elements or electromagnetic induction heating coils

— the controls

The fryer uses hot oil as the heat transfer. There are two basic designs of fry tanks for gas fryers which are the bottom fired design and the tube fired design. Induction fryers also use the tube fry tank design. The fry tank design for electric models has a slightly flat tank bottom with immersion type elements. A "cool zone" is necessary for trapping food particles, minimizing carbonization, prolonging oil life, and reducing flavor transfer. Burners, heating elements, induction heating coils, and gas burners that are commonly used are atmospheric burners which may also be engineered as a power burner and infrared tile burners. Electric heating elements that are commonly used are the flat ribbon style heating elements and alloy stainless steel sheath heating elements.

Two burner range

The range has two gas burners (gas fired models) or electric hot plates (electric models) in a cabinet stand. The range is used to supplement primary cooking equipment. It is used to boiling water and melting butter

Food processor

There are numerous manufacturers of food processors available. Most are compact in design to use less counter space. Use a commercial model not a home-style unit. It is used to;

— texture modification

— food processors are used to speed up the production/ preparation time of some recipe ingredients and or menu items

— food processors dice, slice, grate, shred, and julienne food

— continuous feed units process food quickly with uniform pieces

— labor saving

Ventilation

Cooking equipment in the hot food prduction area can be considered a generator of contaminated air. The quantity of this air is developed by each piece of equipment based on temperature and size of physical cooking area. Properly designed ventilation systems relate to a safe and efficient kitchen. Its uses are;

— air conditioning
— fire safety
— ventilation
— building pressurization
— refrigeration
— air distribution
— foodservice equipment

Holding cabinet

This provides a safe temperature controlled storage of cooked and ready-to-eat food items to meet service demands. All hot foods should be held above 140° F to assure they will not be held in the temperature danger zone. Selection of the holding cabinet with humidity control, for example, is important to maintain food quality. Heated holding cabinets are also known as warming cabinets or heater/proofer cabinets. The unit is an enclosed cabinet designed to hold baking pans or steamtable pans. Their function is to hold prepared food hot for service at a set temperature. Unheated cabinets are used for holding and transporting large quantities of prepared foods not requiring to be held at a set temperature. Holding cabinets come in a variety of styles for both heated and unheated models; mobile or stationary; reach-in, pass-thru, roll-in, or roll-thru. They are made with glass or aluminum doors. Application for heated cabinets include:

— meat
— poultry
— fish
— pizza
— vegetables
— breads, muffins, biscuits, rolls

— proofing bakery dough

Application for unheated cabinets include:

— breads, muffins, biscuits, rolls

— prepared desserts

— catering set-up

Warewashing

The functional area of dish and tray washing is important within the foodservice system. It is here that soiled dishes, glasses, flatware, and trays are visibly cleaned and sanitized. Within the warewashing area there are a number of tasks that must be completed like, scraping, racking, prewash, wash, rinse, sanitize.

Dishmachines

The dishmachine is a major investment and commitment. There are a number of considerations when selecting dishmachine features:

— proper ventilation - otherwise there is excessive humidity and possibly poor/unsafe working conditions

— adequate lighting - allows staff to see broken glass and excess water accumulation

— proper utilities available - inadequate power can damage machine or there can be non-performance

— drain size and location

— walls, floors and ceiling should resist moisture, absorb sound, easy to clean

— booster heater

— capacity based on number of meals

— cleanability

Garbage disposers

The primary function of the garbage disposer is to grind food waste and mix it with water. This mixture is then piped from the operation to the sewer. Commercial garbage disposers are usually specified by horsepower.

Pulper and extractor systems

Some municipalities have ordinance against the use of commercial disposers because of the heavy load on the sewer system. Trash pulpers and compactors can significantly reduce garbage volume. Check local regulations before planning the purchase of a pulper and extractor system. With a pulper, water and waste are mixed together; ground into small particles; piped to extractor; solids are separated from water; water returned to pulper and mixed with fresh water; water reused in grinding process; solids conveyed to conventional trash container. The final product looks like sawdust or ground paper. A pulper and extractor system costs more than conventional disposers.

Trash compactors

Mechanical device designed to compress waste materials so the volume is greatly reduced. Waste is easily transported out of food production area. Trash compactors have many obvious advantages. The garbage can be removed by trucks with a specially designed lift. Kitchens with a large accumulation of glass, boxes, and trash find the compactor to be an advantage.

Role of Grade Manager

Our ancestors developed the practical skills necessary to ensure a relatively steady food supply. This meant learning not only to domesticate animals and raise crops but also to preserve those foods. Fish were brined in seawater and left to dry on the shore, where they either fermented or dried. Meats were hung off the ground and near the fire. This kept them out of the reach of scavenging animals and insects. The smoky bath surrounding them darkened, flavored, dried, and preserved the meats and kept them from spoiling.

Records of various curing methods have been tracked back as far as 3000 B.C.E., when it is believed the Sumerians salted meats as a way to preserve this valuable but perishable food. Historical evidence shows that the Chinese and the Greeks had been producing and consuming salted fish for many years before passing their knowledge on to the Romans.

Food-preservation skills and the necessary ingredients, including salt, sugar, and spices, were greatly valued. Cities such as modern-day Rome and Salzburg were founded near a ready source of salt. As the Romans extended their empire, they conquered lands rich in a variety of resources, including foodstuffs. They brought with them to new territory their own recipes and formulas for a variety of preserved meats, fish, cheeses, wines, and cordials. But the culinary exchange was never in one direction.

Development of the Guilds

Some of these special stored items, such as hams and cheeses, became part of the commerce and trade between towns and states. They were included as dowries and tributes, along with livestock, buildings, servants, and jewels, and served as a kind of currency to acquire other goods. Eventually, rules were established governing how merchants prepared and sold these goods and services, the goal being to prevent monopolies and pricing abuses. The work itself was clearly defined and assigned to various groups known as guilds. The guilds developed training systems for their members, taking them from an apprenticeship to the journeyman stage and, finally, conferring the status of master. Each individual guild was granted a charter that gave its members specific rights.

By the end of the sixteenth century, approximately two dozen guilds were dedicated specifically to food. These fell into two groups-those that provided raw materials and those that provided prepared foods. The guild that prepared and sold cooked items made from the pig was referred to as charcuterie, derived from French root words meaning "cooked flesh." This guild kept the practical work of preserving meats alive and thriving, making bacons, hams, sausages, and pâtés. Numerous strategies were devised to get around restrictions imposed on any given guild, and the charcutières (members of the charcuterie guild) were no exception. One of their tactics led to the development of terrines: Charcutières were not permitted to sell foods baked in pastries.

Making and selling forcemeat loaves baked in pastry-pâté en croûte-would not have been allowed, according to a strict reading of the charcutière's charter. Rather than stop making pâté en croûte, the charcutières baked the forcemeat in an earthenware mold instead of pastry-and so pâtés en terrine were created.

The more essential the food, the more closely it was regulated. The more lucrative a guild's activities, the more likely it was that another guild might be tempted to infringe on it. Each guild fought to protect its individual rights. Several cases were brought before judges to determine if one guild's activities had crossed the line into the work of another's. One such case had a profound impact on the work of today's garde manger. The garde manger (or steward), as well as chefs and cooks, were household employees and as such did not have a formal guild of their own. These workers found their way into restaurants in increasing numbers throughout Europe and the British Isles.

At first, there was no widely recognized structure for kitchen workers. There were no established duties or areas of specialization. It took several years before a serious attempt was made to organize the kitchen workers. Eventually, the brigade system, as recorded by Auguste Escoffier, detailed a logical chain of command that brought order to the unruly working arrangements of his day. We still use the brigade system and refer to the various stations in the kitchen with the names assigned by Escoffier: saucier, rôtisseur, pâtissier, and garde manger. When the guild system was abolished in 1791, some members of the charcutières' guild also joined the ranks of restaurant and hotel kitchen garde manger staffs. Others continued to operate their businesses as before. The work of the charcutière and that of the garde manger have always been closely linked, since both are founded on cold preserved foods. When the term garde manger is used today, it is often understood to include the work of the charcutière as well.

Present Status of Garde Manger

The garde manger, recast in a restaurant setting, has retained its tradition of preparing a variety of preserved and cold foods. It has also expanded its scope to include hors d'ouevre, appetizers, salads, sandwiches, and the accompanying cold sauces and condiments. Garde manger is involved in à la carte service as well as receptions, buffets, and banquets. The techniques required to prepare pâtés, terrines, sausages, and fresh cheeses are the particular domain of the garde manger. To support the techniques and production in the cold kitchen, a wide variety of hot food-preparation techniques that are essential to the foundation of any culinary discipline must be mastered: roasting, poaching, simmering, and sautéing meats, fish, poultry, vegetables, grains, and legumes.

It is precisely because the skills and responsibilities are so broad that many of today's most highly regarded chefs began their careers in the garde manger department as an apprentice or commis. In addition, recent years have seen a rebirth of the more traditional practices of charcuterie and cheese making by purveyors with retail shops and wholesale businesses. Handcrafted foods such as country-style hams, sausages, pâtés, and fresh and aged cheeses are increasingly available to both the restaurant chef and the home cook. Hotels, full-service restaurants, and private clubs may have one or more people working exclusively in the area of garde manger, though its specific name varies from place to place.

Some operations refer to it as the pantry, others may call it the salad station, still others the cold side, and so on. The specific foods for which this station is responsible can include cold sauces and soups, salads, hors d'oeuvre, and canapés. This station may, in some kitchens, supply other stations with particular items. The garde manger, in some operations, also shapes and portions meats, poultry, and fish, adding marinades or stuffings as appropriate.

During service, the garde manger typically plates salads and cold appetizers and may also be responsible for plating desserts. The breakfast, lunch, and brunch menus often rely heavily on the garde manger as well. Cooks and chefs working in banquet and catering operations practice all the same basic cooking skills as the garde manger in an à la carte restaurant. The approach to work is slightly different. This work is so stimulating and challenging that many professionals choose it as their lifelong career path. Here, where the goal is to produce and serve flavorful, attractive food to large numbers of individuals simultaneously, you learn to use the special equipment and cooking techniques of volume production.

The chef not only develops a menu but also does all the planning necessary to come up with scaled recipes, accurate and timely orders for food and other items, and food costs. Presentation is often a significant component of banquets and receptions. Décor, appropriate and effective garnishes for plates, platters, and other food displays, and concerns for food quality and customer safety are considered. The nature of large events often involves a certain level of risk and always calls for the ability to think under pressure and come up with a creative solution to a crisis.

Delicatessens, charcuteries, and shops selling prepared foods of all types offer yet more options for the professional garde manger. Some operations feature handcrafted foods such as cheeses and sausages. Their goods may be sold through a retail shop or exclusively to those in the restaurant trade. Large companies, including hotel and restaurant chains and food manufacturers, look to those with strong skills in garde manger to undertake projects such as the development of a new line of sauces or condiments, spice rubs, or salad blends.

Working Condition

Both employers and schools recognize that formal education on its own is not enough to assure excellence. Garde manger is a practical art. To succeed, you must work hard to develop precise skills. Whether you work for yourself or for someone else, you must make choices about your work carefully. It is tempting to make a decision based on salary, location, or similar immediately tangible reward. However, if you consider each job as an investment in your future, it is far easier to evaluate its long-lasting rewards.

Making wise career choices is complicated, so take the time to evaluate any career move. Develop your own plan for the future as specifically as you can so you can determine the type of establishment and the type of work that will set you up to secure the next level in your career. Look for work environments where every person has a stake in getting things done correctly. When every person has the opportunity to help make decisions and has the tools he or she needs to perform well, everyone succeeds. If you want to do a job well, you need to know the quality standards. Objective evaluations, constructive criticism, and additional training are part of every good working situation.

Work at the entry level includes cleaning and cutting produce, making vinaigrettes and composed platters, and following simple standard recipes under supervision. It is important to ask questions and follow advice, watch carefully what goes on around you, and supplement what you see and hear by reading. Learn the skills necessary for handling special equipment safely and efficiently. Slicers, mixers, grinders, blenders, food processors, thermometers, smokers, sausage stuffers, and salometers are just a few of the specialized pieces of equipment used in the garde manger and smokehouse.

Advanced Level

As your skills improve, you move from entry-level positions into positions of more responsibility. You take on more advanced and challenging work, and you may advance to lead or executive chef. At this level, you have more responsibility for conceiving new menu items, including recipe development, recording standard recipes, and costing and developing and maintaining a budget. You train kitchen and dining room staff in the proper presentation of the new and standard menu items. Banquet and buffet chefs develop menus-both standard and custom-and go through the process of scaling and costing each menu item.

Staffing and scheduling responsibilities for the banquet chef include maintaining and training a relatively large pool of talent, often in conjunction with the dining room manager. Some special aspects of this work include coordinating with other service providers such as florists, musicians, and photographers. Entrepreneurs develop handcrafted specialty items that are produced on both small and large scale. Their work focuses more on the development of a product or product line that is for sale in the retail or wholesale market. They must be concerned with a variety of regulations, certifications, and inspections in order to be sure that foods prepared for sale meet all the necessary legal requirements. Food quality and cost remain as important as ever, and additional business concerns accrue.

Practice of a Profession

Any profession has a great many sides; the culinary vocation is no different. A culinary professional is an artist, a businessperson, a scientist, and a cultural explorer, among other occupations. Acquiring the skills and knowledge necessary to succeed in this profession is a lifelong journey. Employers today look for both experience and education when they hire at virtually all levels higher than entry level. At the most prestigious shops, even entry-level positions may require a degree or some sort of formal training. Employees look for jobs that offer the opportunity to use the skills and education they already possess and, at the same time, to learn new skills.

Formal Education

The increasing emphasis on a formal education goes hand in hand with the emergence of a number of programs dedicated exclusively to the culinary arts. Employers rely on the general and specific skills of the craft taught by schools to establish a common ground of ability. This saves them hours of on-the-job training. The demand for graduates continues to grow each year, and so has the number of programs specializing in the culinary arts. The best education couples as much hands-on practice as course work devoted to product and equipment knowledge. In addition, a well-rounded program provides study in important aspects of culinary arts as a business: customer service, math, food and menu costing, and team skills.

Programs that are recognized in the industry attract high-quality instructors and offer opportunities for students to network, join clubs and organizations, compete, and do advanced studies in an area of specialization. Their graduates receive plenty of hands-on experience and develop confidence and control in all areas of culinary arts. Industry leaders look to graduates of those programs to staff their companies because they bring with them a solid foundation. Even garde manger chefs who already have achieved significant success in their careers take advantage of the many opportunities offered through continuing education. Classes custom-fit to a specific topic give professionals exposure to new techniques and methods and new equipment and ingredients.

Knowledge about food

The ingredients the garde manger uses every day run the gamut from the mundane and utilitarian, such as calves' heads and pigs' feet, to the priceless and exotic-saffron, foie gras, caviar, and truffles. When you know how an ingredient looks, tastes, and reacts with other ingredients, you can use that knowledge to be more creative, more adaptable, and more efficient. At first, you may rely solely on recipes or formulas. As long as everything required by the recipe is on hand, things should work out. Take an extra minute or two to examine each ingredient closely and make note of what it looks likes, how it smells and feels, its shape, and its color.

Classes, workshops, and demonstrations that offer an opportunity to do comparison tastings are excellent learning opportunities. This experience is invaluable, whether your responsibility is to use ingredients appropriately or to buy them in a way that maintains quality and profit. Beyond knowing the color, taste, and cost of an ingredient, today's garde manger typically faces an increasing number of special concerns about the man-ner in which foods are grown, harvested, and processed. A safe and wholesome food supply is a growing concern of both the public and the profession. Topics such as sustainable agriculture, bio-engineering, genetically modified organisms (GMOs), organics, and the support of local and regional growers all factor into the decisions you and your business must make.

Knowledge about instruments

It is true that the foods made by the garde manger and charcutiére are not beyond the technical skill of any good cook, and many individuals enjoy making their own sausages, bacon, and smoked trout. You need not only the correct equipment and ingredients but also the appropriate storage space, one you can keep at the correct temperature and humidity, to produce some items. In addition to being knowledgeable about knives, pots, and pans, the garde manger must be well versed in the use of equipment such as meat slicers and grinders, food processors, smokehouses, brining tubs and salometers, and, for some practitioners, ice-carving tools.

Communication skills

A well-written résumé can sell you to a potential employer. Your own mission statement, if properly worded, keeps you on track and helps you make the best possible career moves. A precise and specific plan for an event can keep it on track and on budget. A thorough and fair interview can unearth the perfect employee or business partner. Each of these activities demands good communication skills. Today's garde manger must communicate using a wider variety of media than ever before, from written memos and letters to e-mails and reports to videoconferencing and interactive learning. A good education program addresses the general and specific communications needs of students and offers courses, workshops, and tutoring or labs in a wide range of communication skills.

Continuing Education

Your education and your experience combine to be the most important source for your professional development. Every career choice or move you make is part of your lifelong education. If you have a long-term plan, you can choose jobs that give you the opportunity to learn new skills and take on greater responsibility as you advance toward your goals. Keeping current with basic skills and new trends is a lifelong task. Once initial training is completed, continuing education is equally important, as the industry is constantly evolving.

Evaluate your career, both as it is right now and as you would like it to be in the future, and then take the appropriate steps to keep on top of the latest infor- mation in the areas about which you are most concerned. Attend classes and workshops, hone your skills in specialized areas, keep up with new ingredients and equipment, learn new management strategies, and strengthen your skills in team building, writing and communication, and marketing and promotion. Some of the courses or seminars you attend can earn you credits (continuing education units, or CEUs). These may be necessary to achieve certain certifications or advancements. Continuing education and professional development programs are available through many colleges and universities, both in traditional and longdistance learning environments.

Not all continuing education occurs in a classroom or over an Internet hookup. Magazines, television programming, newsletters, Web sites, government publications, and books are all excellent sources. Directed travel programs can open up a completely new way of seeing the profession by exposing you to a new cuisine, a new part of the world, a new ingredient, or a new contact.

Knowledge of Networking

The old saying that "it's who you know" has a great deal of truth. The group of professionals you know is known as a network. A solid network is an indispensable tool for the professional and should include members of your profession from as many areas as possible. Knowing someone in a niche not obviously related to your own can turn up unexpected opportunities. Creating a professional network is a task that should be taken seriously. Working with other professionals to share information and knowledge is an important avenue of growth, both professional and personal. Networks can be formal or informal. The way to begin is simply to introduce yourself to others in your field. Have business cards with you when you go to restaurants or trade shows.

Write letters to individuals whose work you have seen and admired. Join professional organizations to expand your network. Well-run groups typically have a variety of meetings and forums to allow members to come in contact with each other. Take advantage of local and national meetings and conventions to learn more about your profession. When you make a good contact, follow up with a phone call or a note. The communication you develop with your peers will keep your own work fresh and contemporary, and an established network makes it much easier for you to find a new job or an employee.

Contests and competitions offer you a chance to stretch yourself. Professional magazines, journals, newsletters, and Web sites have information about contests on the local, national, and international level. Whenever you submit your work to the scrutiny of a panel of judges, you learn. Critical review provides you a means to keep improving in a way that daily production work never can. Practice, research, and the stress of competition exercises your professional muscles, the same way competing in sporting events strengthens an athlete. Even if you are not entered in the competition, attend the judging if possible so you can benefit from the experience.

Garde Manger as Businessperson

Physical assets are the equipment and supplies needed to do business. In the case of a restaurant, these might include food and beverage inventory, tables, chairs, linens, china, flatware, glassware, computers and point-of-sale systems, cash registers, kitchen equipment, cleaning supplies, and ware-washing machines. When we talk about managing physical assets, we are considering how anything you must purchase affects your ability to do business well.

The first step in bringing the expenses associated with your physical assets under control is to know what your expenses actually are. Then you can begin the process of making the adjustments and instituting the control systems that will keep your organization operating at maximum efficiency. One of the biggest expenses for any restaurant is food and beverage costs. You or your purchasing agent will have to work hard to develop and sustain a good purchasing system. Because each operation has different needs, there are no hardand-fast rules, just principles that you will apply to your own situation.

Time Management

It may seem that no matter how hard you work or how much planning you do, the days aren't long enough. Learning new skills, so you can make the best possible use of the time you have, certainly ought to be an ongoing part of your career development. If you look over your operation, you will see where time is wasted. In most operations, the top five time-wasters are

— no clear priorities for tasks,

— poor staff training,

— poor communication,

— poor organization, and
— missing or inadequate tools for accomplishing tasks.

To combat these time-wasters, use the following strategies.

Invest time in reviewing daily operations

Consider the way you, your coworkers, and your staff spend the day. Does everyone have a basic understanding of which tasks are most important? Does everyone know when to begin a particular task in order to bring it to completion on time? It can be an eye-opening experience to take a hard look at where the workday goes. Once you see that you and your staff need to walk too far to gather basic items or that the person who washes the dishes is sitting idle for the first two hours of the shift, you can take steps to rectify the problem. You can try to reorganize storage space. You may decide to train the dishwasher to do some prep work, or you can rewrite the schedule so the shift begins two hours later. Until you are objective about what needs to be done, and in what order, you can't begin the process of saving time.

Invest time in training others

If you expect someone to do a job properly, take enough time to explain the task carefully. Walk yourself and your staff through the jobs that must be done, and be sure everyone understands how to do the work, where to find necessary items, how far each person's responsibility extends, and what to do in case a question or emergency comes up. Give your staff the yardsticks they need to evaluate the job and determine if they have done what was requested, in the appropriate fashion, and on time. If you don't invest this time up front, you may find yourself squandering precious time following your workers around, picking up the slack, and handling work that shouldn't be taking up your day.

Learn to communicate clearly

Whether you are training a new employee, introducing a new menu item, or ordering a piece of equipment, clear communication is important. Be specific, use the most concise language you can, and be as brief as possible without leaving out necessary information. If a task is handled by a number of people, be sure to write it out, from the first step to the last. Encourage people to ask questions if they don't understand you. If you need help learning communication skills, consider taking a workshop or seminar to strengthen your weak areas.

Take steps to create an orderly work environment

If you have to dig through five shelves to find the lid to the storage container you just put the stock in, you aren't using your time wisely. Planning work areas carefully, thinking about all the tools, ingredients, and equipment you need for preparation and throughout service, and grouping like activities are all techniques that can help you organize your work better. Poor placement of large and small tools is a great time-waster. Use adequate, easy-to-access storage space for common items such as whips, spoons, ladles, and tongs. Electrical outlets for small equipment ought to be within reach of everyone. While you may be forced to work within the limits of your existing floor plan, be on the lookout for products and storage strategies that can turn a bad arrangement into one that works smoothly.

Purchase, replace, and maintain all necessary tools

A well-equipped kitchen has enough of all the tools necessary to prepare every item on the menu. If you can't purchase new equipment, then think about restructuring the menu to even out the workload. If you can't remove a menu item, then invest in the tools you need to prevent a slowdown during service.

Information Management

The garde manger is part of the much larger culinary world. Read about all areas that might affect your career and your industry: business and economics, arts and entertainment, society and politics. Popular culture has a curious way of influencing your work. Your customers and clients do not live in a vacuum, and neither should you.

There are numerous print and online sources devoted to the specifics of new or unusual ingredients, unfamiliar dishes or equipment, and more. Information gathering can become a full-time task on its own. To make use of available information, you must be able to analyze and evaluate carefully to sift out the important material from useless data and use all sorts of media and all sorts of technology effectively. Learn more about the profession's history, not just because it is interesting, but because it gives relevance and ballast to the decisions you make.

People Management

Restaurant operations rely directly on the work and dedication of a number of people, from executives and administrators to line cooks, wait staff, and maintenance and cleaning staff. No matter how large or small your staff is, your ability to engage all your workers in a team effort is one of the major factors in determining whether you will succeed or not. Most people prefer to work in an environment where everyone can make a distinct and measurable contribution. The first task in creating such an environment is a properly written job description. Training is another key component. To do a job well, the employee needs to know the quality standards and have those standards consistently reinforced with clear, objective evaluations, feedback, constructive criticism, and, when necessary, additional training or disciplinary measures.

Everyone has the right to work in an environment that is free of physical hazards. This means that, as an employer, you must provide a workspace that is well lit, properly ventilated, and free of obvious dangers, such as improperly maintained equipment. Employees must have access to potable water and bathroom facilities. Beyond this bare minimum, you may offer a locker room, a laundry facility that provides clean uniforms and aprons, or other such amenities. Workers' compensation, unemployment insurance, and disability insurance are also your responsibility.

You are required to make all legal deductions from an employee's paycheck and to report all earnings properly to state and federal agencies. Liability insurance must be kept up to date and at an adequate level. Employers may choose to offer additional forms of assistance as part of an employee benefits package. Life insurance, medical and dental insurance, assistance with dependent care, adult literacy training, and enrollment in and support for those enrolled in substance abuse programs are examples of the support an employer can provide employees.

6

Menu Planning

Before beginning to use one of the menu planning systems, review the basic principles of good menu planning. No matter which menu planning system is chosen, begin by putting foods together using a plan that will result in a menu that is nutritious and appeals to students. The following basic concepts are important for success.

The menu is...

— the driving force of the operation.

— a management tool that controls program functions.

— dependent on the knowledge and skills of the planner.

— more efficient with a cycle.

— improved by offering a variety of foods.

— focused on nutritional needs of children.

Driving Force

The menu is the basis for all food service program activity. It drives the planning, purchasing, production, service, cleanup and nutrition education functions. Menu planning also provides an opportunity to involve children, parents and teachers in the programs.

The menu must be planned to be in compliance with federal regulations and program requirements. The menu determines the nutrient content of the meal and the meal acceptability. The acceptability influences the participation rate. The menu also determines the food cost, and its complexity affects labor costs. The menu controls what purchasing and food production must be done to produce the meal. The food production plan determines how the work is scheduled and what equipment is used. Employee training needs are determined by what foods are on the menu and how they are prepared and served. The menu plan sets the food service program in motion and controls many of its functions.

The quality of the meals and the success of the program depend on the knowledge and skills of the menu planner. Menu planners need to know...

— Program purposes and goals, requirements and recommendations.

— Students' food preferences.

— Food costs and the amount of money available.

— Foods available for the menu planning period.

— The availability and experience of personnel.

— Kitchen layout, type and capacity of equipment.

— Food preparation and work scheduling.

— Food merchandising for customer acceptance.

Cycle Menus

Using cycle menus for any of the menu planning systems will save time and increase efficiency. A cycle menu will not only save time for the grade or age group for which it is planned, but with changes in portion size the cycle may be adjusted to use for another grade or age group. The menu planner may select a time frame for a cycle menu. The time frame can be one, four or any other number of weeks that works for the individual operation. To avoid repetition in a cycle menu, changes may be made to incorporate seasonal variations and special events into the cycle. This is still easier than starting over every month. In NuMenus and Assisted NuMenus, there are steps to take before actually completing the menu plan and analyzing the menu. These steps include entering in the approved nutrient analysis software the recipes and vendor nutrient analyses for menu items, as well as the RDA age or grade group.

In Food-Based Menus, there are also steps to do before completing the planned menu. Calculations must be done to ensure that the correct quantities of food to be purchased and prepared are known for each food component and item. Using a cycle menu can save time on these preliminary steps from month to month, leaving more time for training staff and promoting healthful food choices.

Variety in the menu encourages consumption of healthy foods. Offer students a selection of foods and types of milk from which to choose. To increase food consumption and participation in schools that do not offer choices each day, no one meat or form of meat should be served more than three times in a week. "Form of meat" refers to ground, sliced, pieces, etc. For planning purposes, make a list of all of the school's recipes, vendor products and other program foods. Use this list to avoid continually repeating the same items rather than offering all of the available items occasionally. Lunch, breakfast and special meal menus should be coordinated in order to avoid serving the same food at different meals.

Nutritional Needs of Children

Adjust Portion Size

Adjust the portion size for the various ages or grade levels of children, whether using NuMenus or Food-Based Menus. As mentioned under cycle menus, if a basic cycle menu for one age group has been planned, the portion sizes may be changed to meet the nutritional needs and/or program requirements of another age group.

Minimize Waste

When portion sizes are adjusted, food waste with younger students will be minimized. Portion sizes that are too large discourage young children from eating and also may lead to overeating.

Steps in Menu Planning

Collect menu resources: There are many menu resources available:

Recipe Resources:

- USDA Quantity Recipes for School Meals
- USDA Tool Kit for Healthy School Meals
- School recipe file(s)

Other Resources:

- School calendar with important dates
- Student participation records
- Production and inventory records
- KSDE Menu Planning for Healthy School Meals
- Food trade journals

Select the grade or age group: First select the grade or age group based on the grades in the school or group of schools for which the menu is planned. The grade or age group selected will determine the type of menu items and the appropriate portion size. Later, the portion sizes may be adjusted to other grades or age groups, or as needed to meet program requirements.

Determine number of choices: Determine the number of choices that will be offered for each food item or component for Food-Based and Traditional, or each menu item for NuMenus. Providing choices and variety is an important concept. The number of choices offered in each category depends on the operation. Look for a balance in cost, nutrients and equipment usage, as well as the labor needed to prepare each item. Start by adding one or more entrées and then adding a selection of side dishes that complement the entrée choices. This approach is appropriate for any of the menu planning systems.

Offering choices has the added benefit of allowing the introduction of new foods without the usual drop in participation. Letting students "take a taste" of a new menu offering is the ideal way to introduce students to a wider variety of menu selections. Offering choices at sites where Offer versus Serve is in place encourages students to select foods they intend to eat. Extensive choices do not need to be offered. Even offering just two choices gives students the opportunity to express their individual preference and increases the likelihood that a full meal will be selected.

Evaluate the starting point: Start by looking at the current menus, purchased products, recipes and preparation techniques. For NuMenus, analyze and evaluate how well the menus meet the Nutrient Standards.

Determine time period: The nutrition goals for all three menu planning systems are set for a period of one school week, which is most often five days. The menu planner should plan menus by the week. For the purpose of nutrient analysis, combine any week with less than three consecutive days with either the prior week or the coming week. For Food-Based Menus, the menu planner needs to keep in mind the minimum servings for certain food items per 5-day week. For weeks with less than 5 days, the weekly requirement will need to be adjusted.

Select the entrée for lunch or the main course for breakfast: The meat or meat-alternate usually sets the scene for the rest of the lunch menu and sometimes for the breakfast menu.

It may be the determining factor as to whether students elect to eat that day. Therefore, careful planning of the meat or meat-alternate can improve participation. The entrée (or main course) for lunch is usually a meat/meatalternate; but could also include grains/breads or vegetables/fruits. Even though breakfast does not include an entrée, the main course may be comprised of grains/breads and/or meat/meat-alternates.

Select the other menu item(s): Select other menu items that complement the entrée or main course.

a) *Contrast*: This is the opportunity to add color and texture as other menu items are planned for the meal. To add color, use bright fruits and vegetables or a colorful dessert item. To add texture, use crisp, firm foods.

b) *Balance*: Balance in "weight" and "flavor" can also be achieved as the other menu items are added. If the first menu item planned is heavy, plan a light vegetable or a dessert such as fresh fruit. If the entrée or main course is light, add a higher calorie food such as a healthy baked dessert, a cooked vegetable (such as potatoes or corn), or a bagel at breakfast. To balance flavor, use a combination of mild and strong flavored foods. Too many foods with strong flavors in the same meal may result in an unacceptable meal.

c) *Variety in Shapes and Sizes*: Another opportunity for creating appealing menus is with shapes and sizes. Consider a meal with fish sticks, oven-baked Frenchfries, carrot sticks and a banana. All of the above foods have a similar shape. Presenting foods in several different shapes appeals to children. For example, the following menu offers good variety of shape and size: baked chicken leg, potato rounds, carrot sticks, watermelon wedge.

d) *Color*: The last, but perhaps one of the most important considerations, is color. Consider the menu with the fish sticks. Not only was every menu item the same shape, they were also in the same color family. It helps to use at least two colorful foods in each menu. Vegetables and fruits are a natural way to add eye appeal. It helps to add a bright colored food to a menu with little or no color. For example, add a slice of tomato to a potato salad or put a fresh grape or strawberry on a dish of diced pears or peaches. A dash of cinnamon or paprika can be used to achieve the same effect.

Provide fluid milk choices: Low-fat milk options should be available every day.

Meet nutrition goals: The challenge of meeting the nutrition goals, including providing the calories and key nutrients of the Recommended Dietary Allowances and meeting the Dietary Guidelines for Americans should also be considered as menus are planned. The ultimate goal is to reduce fat, saturated fat and cholesterol while maintaining calories and nutrient levels.

Evaluate: Menu planners using NuMenus will analyze nutrient content and adjust their menus before the planning is complete. Menu planners using Food-Based Menus will evaluate their menus to ensure that the correct food components and servings of food items are included. Nutrients provided by the menus will be analyzed and evaluated during a SMI review conducted by KSDE school food service consultants.

The following requirements must also be considered in the menu planning system.

1. School site menu versus centralized menu
2. Standardized recipes and preparation techniques
3. Processed foods analysis
4. Condiments
5. Projected servings

School Site Menu and Centralized Menus

If all schools in a grade or age group follow the same centralized menu, the menu planner may plan and analyze just one menu for each group. If each school is allowed to select its own menu, then menu planning and analysis must be done for each school site. If schools have traditionally planned their own menus, an option might be to plan part of the menu centrally and then let each school add to the basic menu. For example, all schools could serve the same entrée(s), but offer different side dishes. This makes it simpler to control fat content since the menu item with the highest fat is often the entrée. It also allows flexibility within local schools, but saves data entry time over totally individual menus. When projecting servings and portion sizes for each menu item and condiment, count all schools for a centralized menu or each individual school for individualized school site menus.

Recipes and Preparation Techniques

For every menu item, there must be in place a standardized recipe and a standardized preparation technique. Menu planners should identify the standardized recipe and preparation technique as they plan the menu to ensure that the food planned is the same as the food served.

Processed Foods

- Child nutrition labels are required for pre-prepared items.
- Nutrition labels are required for pre-prepared items.

Condiments

Condiments are optional items and are not counted as being a menu item. Examples of condiments include:

- catsup

— mustard
— jelly
— salad dressing
— gravy
— chopped lettuce or tomato served as a garnish
— pickles served as a garnish

Even though condiments are not menu items, menu planners must include the projected portion sizes and servings of all condiments in their menu production record. This information is needed because all foods (except foods of minimal nutritional value not included in a menu item) count toward the nutrient analysis when it is time to analyze and adjust the planned menu. In the past, these foods were extras and were not accounted for in the meal pattern. But now they must be included in the nutrient analysis.

Planned Servings

Menus are analyzed for nutrients based on the planned servings of each menu item and condiment. By giving more weight to menu items that are selected more often, the analysis will give an accurate picture of how well the menu is meeting the nutrition goals. It will be easy to see which menu items could be modified or replaced to make a large impact on the nutritional value of the meal. The menu planner will have a view of the nutritional content of the meal as planned for consumption by the students. Menu planners need access to prior menu production records in order to project future servings for the menu items and condiments.

7

Catering Management

"Services" in the context of a foodservice operation refer to the opportunities management makes available to the customer to purchase food, beverage, entertainment, and ancillary services. Services include:

1. Table foodservices
2. Packaged take-out foodservices
3. Beverage services
4. Entertainment services
5. Business meeting services
6. Conference and convention services
7. Contract feeding services
8. Off-premise foodservices
9. Home-replacement foodservices

Full-Service Restaurants

Full-service restaurants have the opportunity to offer a variety of catering services to their customers. Before any decision is made to offer these services, however, six important factors should be considered:

1. Location
2. Customer profile

3. Restaurant style or concept
4. Staffing capabilities
5. Restaurant physical layout
6. Cuisine and menu offerings

Location

The proximity of the restaurant to office complexes and centralized business areas will help to establish whether its catering service will be focused on business or social marketing efforts. Businesses in the twenty-first century are spread from urban centers to suburban locations in office parks. Central urban locations offer a concentrated market for both office delivery and take-out. A significant factor in developing the market for business catering is that service is generally required during the business week, leaving weekend periods free to service social business. In addition, locations such as museums, concert halls, and historical sites offer interesting venues to catered functions for both local businesses and conventions.

Both urban and suburban restaurants can successfully develop social catering business. Suburban locations are generally more appropriate for social catering to private homes, clubs, churches, and other facilities. Delivery to urban locations can pose security and logistical problems, creating additional costs for transportation and service labor. Population density also affects the volume of anticipated catering business. Restaurants situated in rural areas with low population density cannot expect immediate high volumes of catering business.

Areas of high-density population yield a variety of catering opportunities that steadily increase in volume through referrals and reputation. The location of the physical restaurant building plays a role in the type of catering services to be offered. Storage facilities, expansion possibilities, and access to major transportation routes are factors important to catering service production.

Customer Pool

Restaurants have the advantage of a built-in customer pool to whom they can market in-house and off-premise catering services. In addition, the attraction of being associated with a restaurant's reputation will help to expand the possible market to include new business and social clients. The market profile should classify customers as business or social catering clients, designated by income bracket. In addition the range of activities for which each customer pool will need catering services should be researched as thoroughly as possible. This will help in developing package programs along with potential menu programs and accompanying pricing concepts.

Style of a Restaurant

The style, concept, and/or theme of the restaurant should be taken into consideration when planning potential catering services. Off-premise catering services do not necessarily have to blend with the facilities offered by the restaurant. On-premise catering services should, however, be designed to function within the restaurant facilities.

Facilities

Restaurants facilities are a major factor in providing on-premise catering. The ratio of catering functions to restaurant services that can be handled at a given time is dependent cn the size and flexibility of the physical plant. Small private parties are often incorporated into the general dining room setting. Large parties must, however, be given facilities that are separated from the general public. Flexible facilities such as these allow a restaurant to maximize revenues.

Many restaurants that offer in-house catering schedule large parties, such as weddings, anniversaries, luncheons, and dinners, on days and times when the restaurant is not otherwise open. Often catering business must be refused because sufficient on-premise facilities are not available. When management is continually turning away catering business, a decision will need to be made as to how the restaurant will balance is future development of catering versus full-service dining services. Kitchen facilities play a major part in determining when and how catering service demands can be met. Kitchen equipment must be flexible, allowing for volume production to take place simultaneously with à la carte restaurant service.

The kitchen cooking load and holding capacity of ovens and auxiliary equipment is important to determine if a kitchen is to be used to its full capacity during busy times. Storage and refrigeration facilities determine the amount of food products available at any given time. The cost of waste from food spoilage due to lack of refrigeration and freezer space could dilute the profit from additional catering business.

Cuisine and Menu

The primary cuisine and menu offerings of a restaurant constitute one of the most important considerations for on-premise catering. Off-premise catering services do not necessarily need to be the same as those offered at the restaurant for full-service dining. Purchasing and production requirements are crucial to the successful development of catering services. On-premise catering should offer menu items that duplicate the established menu as closely as possible in order to enhance production capabilities. Surrounding items, such as vegetables and starches, are the most effective area of the menu to duplicate.

Kitchen production is more efficient when the number of surrounding menu items is limited. Profitable and effective purchasing for catering functions requires that the ingredients for menu items be the same or similar to those on the restaurant menu. This allows the purchasing agent to place orders for maximum volume pricing and quality.

Off-premise Catering

The decision to participate in off-premise catering requires a thorough review of a restaurant's resources. A successful in-house catering program can lead management into perceiving that taking on the challenge of off-premise catering will not cause operational difficulties. Several problems must be anticipated regarding the off-premise facilities in which the final production and service for functions takes place. During in-house catering functions, service staff can often work between private parties and the dining room, filling in where needed, while the kitchen staff functions as usual, having planned ahead to handle the increased production load.

In order to service off-premise functions, however, a separate waitstaff needs to be scheduled. In addition, at least one member of the kitchen production staff will be assigned to the function and therefore will be unavailable to the restaurant kitchen for the time period involved for the off-premise function. Equipment as well as food will need to be transported both to and from the off-premise function location. Hot food and cold food must be kept at specific temperatures to prevent food spoilage. A refrigerator truck may be necessary at certain times of the year. Transportation routes and problems involving traffic congestion need to be considered in the timing of deliveries.

Subcontractors may be necessary for equipment such as tables, chairs, dishes, glassware, linens, dance floors, and tents. The extent to which a restaurant commits to a full-service catering business determines the amount of extra effort required to handle the increased business. Planning for offpremise catering requires attention to detail. Contingency plans for bad weather, delays in food transportation, and the failure of on-site equipment to function properly should be established in advance so that problems can be dealt with expediently as they arise.

Additional costs are incurred when solutions to these situations are not planned for in advance. Contract prices must include these contingencies or management will find that overall costs are higher than the fees charged to the customer for food and services.

Take-out Service

Take-out service is the most profitable way for restaurants and catering services to increase revenues without increasing costs. Because take-out service does not require extra seating, it generally will not require an operation to expand its facilities beyond accommodating takeout customers with a pickup and waiting area. Kitchen production can plan and schedule to handle large increases in volume for short time periods. In addition, take-out does not require the additional expenses incurred by table service for glassware, linens, flowers, menus, and entertainment. Disposable dishes, plastic flatware, and a carrying container are the basic requirements for take-out service. Successful take-out service does, however, require planning to assure that customers receive menu items in satisfactory condition. As with the catering menu, the selection of take-out menu items should be drawn from the restaurant menu in order to maximize purchasing and production efforts.

Hotel Food and Beverage Department

The hotel food and beverage department provides food-related guest services throughout a hotel, conference, or resort property. Outlets for food and beverage services can include the following:

1. Full-service restaurant
2. Coffee shop
3. Catering facilities
4. Room service
5. Recreational areas
6. Lobby area bars
7. Food market or delicatessen

Food items for these combined foodservice areas are provided from a central kitchen, with the exception of large hotel facilities where satellite kitchens provide auxiliary production and service areas. An executive chef supervises production for all areas, consulting with the manager of each foodservice area. Of the seven food service areas, catering affords the possibility for the greatest profit, in addition to providing much-needed cash flow during periods when room sales are slow. Hotel catering services are usually classified as:

1. Business
2. Convention/conference
3. Social

The volume of catering service available to a hotel is based on the size and number of facilities that can be used for private functions as well as the availability of production and service staff and related equipment. The primary catering market for a hotel is based on three factors:

1. Location
2. Hotel facilities
3. Customer profile

Location

The location of a hotel determines the demand for in-house

catering. Inhouse catering is defined as guest-related foodservice functions associated with business meetings, conferences, conventions, and social concerns. A hotel with a remote location focuses its catering efforts on the business booked into the hotel. Resort locations often provide additional opportunities for catering services at sports facilities and landscape venues throughout the property.

A hotel with a suburban or downtown location can expect to develop a large volume of outside catering business, both business-related and social in addition to conferences and conventions. The Chicago Hilton, for example, located in downtown Chicago, breaks down its annual catering business into the following percentages:

— 65% conference/convention

— 35% outside social

The amount of combined catering business that can be handled at any one time in a hotel is based primarily on the number and size of the facilities available. The major function rooms are usually referred to as the ballroom areas. These are complemented by a series of meeting rooms, some adjacent to the main ballroom area and some located in other areas of the hotel.

Customer Profile

The customer profile for hotel catering services is made up of both business and social clientele. Customers who use the hotel as a location for a twoor three-day meeting consider meal functions as part of the total meeting package. This arrangement not only enables customers to work within a budget that covers the entire project, but also fits with an approach to planning catering functions that considers the success of the overall meeting rather than each individual function. The catering manager is generally given the responsibility of creating a menu plan for the entire program along with theme functions and special events.

The target market for hotel catering functions is a customer who reflects the social, business, and economic profile of the community. Hotel catering services attract a wide base of customers with a broad range of requests. Customers who plan business functions on a regular basis are familiar with the catering staff and services of the hotel. Their needs are often quickly identified and provided for. Customers who plan social functions on a one-time basis, however, need more personal guidance throughout much of the planning stage. Pricing is a primary concern of this customer group.

Off-premise Service

Hotel food and beverage departments are often requested to handle the arrangements for functions to be held away from the hotel property. If a company wants to retain as much revenue as possible from business, convention, and social customers, an off-premise catering policy should be developed. Often, a hotel will have the facilities in place to offer a limited menu of takeout items. Items have been selected from the hotel dining room menu to offer to the general public on a take-away basis. Hotels have the advantage over many private catering companies of being able to use their own equipment. In addition, kitchen and storage facilities are available to handle large volumes of both food and beverage as well as equipment. If the volume of on-property business requires the use of hotel equipment, rentals can cover the additional needs, with costs being charged back to the customer.

Development

The development of foodservice concepts for hotels has evolved in recent years from the traditional view that considered the food and beverage department as a necessary evil to the modern idea that the food and beverage department is an important profit center. Some large hotels have food and beverage sales of over $35 million per year, an amount that exceeds room sales and creates in management a high expectation of profit from these two departments. The Hilton Hotel in Atlanta, Georgia, has developed a concept for its first-class rooftop restaurant that goes beyond the idea of a foodservice facility as a profit center.

The decor is exquisite, the food is served with flair and showmanship, and the entire theme captures the imagination of the city's residents. The concept was developed with such success that the hotel's own guests had great difficulty getting reservations to dine. A hotel restaurant so overcrowded that it could not serve the guests of the hotel would have been unthinkable in earlier days of hotel keeping in the United States. Nikolai's Roof is an excellent example of the execution of a hotel dining concept that complements the hotel itself as well as drawing a significant number of guests from the community. Hotel managers have known for many years that hotel restaurants must have certain desirable features if they are to be successful. These features include:

— Availability of parking
— Unique theme or decor (differing from the decor of the hotel itself)
— Strong promotion to the community
— A menu and a method of service that are distinctive

The developers of hotel properties, and in some cases hotel chains, have used outside foodservice facilities and interior design consultants to create unique specialty restaurants that can be successfully marketed to both hotel guests and the community.

Catering Halls

A catering hall is a facility dedicated to private parties with an on-site production kitchen and staff. These facilities can offer a wide range of both inhouse and off-premise catering services. The major factors that influence the market for a catering hall are:

1. Style or concept
2. Facilities
3. Customer profile

Catering halls offer customers a self-contained private function space independent of a hotel or restaurant. Many catering halls specialize in social functions such as weddings and plan the design of the landscaping and building to accent such functions. This design is also often aimed at a particular segment of the customer market, depending on location of the hall and the density and ethnic background of the surrounding population. Catering halls can also specialize in large functions, providing space for groups that cannot normally be accommodated in other facilities. Such halls are generally designed to host a wide range of social and business functions and a variety of special event themes.

Facilities

The flexibility, and often the originality, of function spaces are important to the success of a catering hall. Customers will travel considerable distances to social functions held in facilities that offer a unique setting or experience. Wedding receptions are often planned in locations over an hour away from the site of the ceremony so that the bride and groom can have a specific setting for their reception. Catering halls can provide the setting for special event themes ranging from futuristic to country western, offering menu and entertainment packages to match.

Customer Profile

The customer target market for catering halls, like hotels,

is based on the social, business, and economic profile of the community. Catering halls, also like hotels, are available in a variety of price venues, generally appealing to customers who are looking for menu and function packages within a specific range of prices. Identifying a customer group toward which to target marketing efforts includes surveying the competition, establishing a price range, and determining specific geographic areas of the community in which potential customers live based on demographic information such as annual income, family size, and age.

Off-premise Catering

Catering halls have a great deal of flexibility regarding the type of business that they can solicit. The amount of off-premise catering that can be handled by an operation is limited only by the ability to service it. Location, transportation routes, and population density play a major role in determining this segment of the business. The ability to transport food and equipment and provide service and production staff are other limiting factors.

Private Caterers

Independent caterers are private businesses offering catering services to the general public. These businesses operate with or without permanent facilities of their own in which to hold functions. Those caterers who operate without a formal facility must arrange for kitchen production and storage space. Kitchen space can be rented from schools, churches, senior citizen centers, and other institutional facilities with commercial refrigeration and production space. Storage space can be arranged for in self-storage warehouse rental facilities that combine short-term leases with flexible storage space and 24-hour access. Some catering companies have their own production facilities and enough warehouse space to handle large volumes of business.

Chairs, floor covering, tables, and the table setup necessary for large functions are either sourced from a catering company's own warehouse or contracted through a local rental supplier. Many independent caterers focus on off-premise catering, maintaining only a small catering hall with limited facilities to house their production and storage needs. Their focus is on both business and social functions within a wide geographical area. In order to successfully cater to as many functions as possible simultaneously, a catering company has to be extremely well organized. Although equipment needs, staffing, and transportation are major concerns, the primary product, food, is the most important issue.

Duplication of Menu Items

The key to successfully providing catering services to a number of functions simultaneously is the duplication of as many of the menu items as possible in order to minimize kitchen production. All surrounding items, salads, and desserts can be standardized. The main course selection should be limited to three or four items on any given day if possible. Menu items should be chosen for their ability to be precooked for completion in another location. Foods should have excellent holding properties and retain heat.

Dessert #2 is a chocolate dessert shell filled with frozen yogurt that can be preassembled or purchased frozen and transported to function locations within reasonable proximity to the main function kitchen. The gateau chocolate is contracted from an outside baking source, as are the breads. Whipped cream is served with the cake using a high-quality product from aerosol containers.

The main course offers a choice of two meats, one poultry, and one fish. All four of these items fulfill the requirements for preproduction and holding properties. Two vegetables selections are offered, one appropriate for the meat dishes and one better-suited for the poultry and fish, but acceptable for any of the main course selections. Standardization procedures such as these minimize the problems of serving multiple parties in off-premise settings and ensure that food presentation is of the best possible quality.

Private Clubs

Private clubs offer a self-contained facility that operates both full-service dining rooms and private function space along with a variety of food and beverage outlets. Private clubs are generally dependent on their membership for both dining room and catering business. Functioning as nonprofit organizations, clubs are, in many areas, prohibited by law from accepting or soliciting business from nonmembers. If such is the case private clubs may, however, cater functions that are sponsored by members and attended by nonmembers, allowing them to service both social and business activities. Factors that influence the success of restaurant and hotel catering services, such as location, facilities, and customer profile, also affect private clubs.

Off-premise catering services can also be offered to private club members. Private clubs may also contract out their foodservice operation to a contract foodservice operator. An example of this type of arrangement is The Colonial Inn, Concord, Massachusetts, and the local country club. The management of The Colonial Inn turned to the local country club as a source of both kitchen production space and private function rooms. The Inn itself, a historic property located in the middle of a crowded town, has no available adjacent property on which to develop additional catering facilities. As a vehicle to increase their food and beverage revenues without incurring capital expenditures, the catering management team had identified a market for off-premise catering in the suburban Boston area adjacent to Concord. Without additional kitchen production space, however, the project could not go forward.

In addition, during heavy business periods throughout the year, the Inn could not accommodate all of the requests that they received for private functions on property. Through an arrangement with the local country club, management now not only has a satellite kitchen in a nearby facility but can also lease function space from the club on an as available/as needed basis. This alliance creates revenues and capitalizes on the needs of The Colonial Inn and the facilities of the club.

Location

Clubs are broken down into two general classifications: private clubs and country clubs. Private clubs are generally situated in the downtown areas of larger towns and cities. The club facility is used primarily as a location for meetings, private dining, social and business functions. Country clubs require a suburban location and usually promote a combination of golf, tennis, and/ or boating as their primary recreational facilities. Private dining and catering facilities are located in the clubhouse. Acceptability of location is based primarily on general location and accessibility to major transportation routes. Downtown city locations can create problems regarding parking and security for people traveling into the city.

Club Facilities

Club facilities can greatly influence the volume of catering business. Clubs offering spectacular settings, access to garden areas for outdoor functions, or well designed interiors will easily attract social business.

Off-premise Catering

A club's ability to accept off-premise catering function business is controlled by the same laws and regulations as the solicitation of nonmember business. Beyond these restrictions, the same considerations for restaurants and hotels concerning transportation, staffing, production, and service apply.

Contract Feeding

Contract feeding companies provide institutions such as hospitals and schools, as well as businesses, with in-house meal programs designed to meet specific needs. Food production and service is contracted for long-term periods with designated budget restrictions. The contract feeding segment of the foodservice industry has been steadily growing since the late 1980s. Major corporations such as Marriott International created large contract feeding companies to service a wide range of customers. In the late 1990s a separate company, Host Marriott, was created and now operates independently of Marriott International, serving institutional customers internationally.

As food costs have increased and available labor decreased, institutions and businesses have turned to outside contractors to operate their foodservice programs. In most cases, these programs function as a service to either patients, clients, or employees; they are subsidized benefits, not profit making operations. By using a professional contract foodservice operator to handle this area of operations, businesses find that the quality of service increases significantly along with client and employee satisfaction. The major factors that can influence the success of catering services within a contract feeding location are:

1. Facilities
2. Customer profile

Facilities

Contract feeding can be offered in a wide variety of dining room and catering facilities. The type and number of facilities that can be used for function spaces within a business or institutional setting will help to determine the volume of in-house catering. Catering functions in these settings are usually business related, but open area spaces can be used by private groups for social functions.

Customer Profile

Identifying the customer profile for large and diverse groups of people in a business situation is initially done by job classification level. Private dining room service and business meeting menu selections should be based on the expectations associated with the job levels of the people who attend these functions. The types of food products that will be popular in a cafeteria area depend on the profile of the employee population group. The proximity of the business to an urban center and the general education level of the customer group, along with local and regional food preferences and subsidy budgets, determine the range of food products offered.

Off-Premise Catering

Contract feeding companies working for a business or institution are often called upon to produce foodservice functions in off-premise locations.

Charcuteries and Delicatessens

Charcuteries are food stores that offer take-out foodservice along with gourmet food products. Many of these products are used as ingredients in the preparation of the food line. In some cases a small seating area will be offered for eat-in customers. The trend for home-replacement food is significant to this type of operation. When customers want something beyond fast food for a take-home meal, they turn to establishments that can provide meals that they would like to have prepared for themselves. Home-replacement food is being provided in a number of types of foodservice facilities, including upscale supermarkets. This type of foodservice operation usually specializes in a regional or national cuisine. Menu items range from salads and sandwiches to fully prepared meals. The success of a charcuterie depends on two major factors: location and customer profile.

Location

Charcuteries should be located where the customer profile supports the level of cuisine offered. Pricing for this type of foodservice operation is generally higher than that of delicatessens and supermarket preprepared and packaged food items; hence it requires a customer base willing to pay for gourmet food products. Smaller independent caterers often operate out of this type of retail outlet, using the facility for both production and storage in addition to a means of advertising and promotion. Delicatessens are food stores that provide a variety of food products and offer both eat-in and take-out food service. The location should be reasonably close to the targeted customer base. Menu item offerings range from sandwiches, salads, and pizza to preprepared meals-to-go. Take-out foodservice of preprepared foods is a major business segment for this type of foodservice operation.

Customer Profile

The profile of customers for a charcuterie/delicatessen operation is drawn largely from the demographics of the population in the immediate area of the business. Primary customers can be identified as either employees of adjacent businesses or local area residents. A customer profile is determined for each of the two primary customer groups. For example, the twice a week business customer, with an average check of five to six dollars, could be in the mid-20s to 40s age group and low to middle income bracket. The local resident profile, with an average check of twelve dollars or more two times a week, may be in the mid-30s to 50s age group, middle to upper income bracket. Further investigation may show that the business employee is a morning coffee and lunch customer while the local resident is an evening customer for home-replacement meals.

Off-Premise Catering Management

Off-premise catering can mean serving thousands of box lunches to a group of conventioneers, serving an elegant dinner for two aboard a luxury yacht, or providing food, staff, and equipment for an upscale fundraiser with hundreds of guests. In most cases there is no existing kitchen facility at the location where the food is served. Caterers provide single-event foodservice, but not all caterers are created equal. They generally fall into one of three categories:

— Party food caterers supply only the food for an event. They drop off cold foods and leave any last-minute preparation, plus service and cleanup, to others.

— Hot buffet caterers provide hot foods that are delivered from their commissaries in insulated containers. They sometimes provide serving personnel at an additional charge.

— Full-service caterers not only provide food, but frequently cook it to order on-site. They also provide service personnel at the event, plus all the necessary food-related equipment—china, glassware, flatware, tables and chairs, tents, and so forth. They can arrange for other services, like décor and music, as well. In short, a full-service caterer can plan an entire event, not just the food for it.

Off-premise caterers meet the needs of all market segments, from the low-budget customer who looks for the greatest quantity and quality for the least amount of money, to the upscale client with an unlimited budget who wants the highest level of service, the ultimate in food quality, and the finest in appointments—crystal stemware, silver-plated flatware, and luxurious linens. Between these two extremes is the midscale market segment, which requires more quality than the low-budget sector, but less than the upscale.

Off-premise catering is an art and a science. The art is creating foods and moods, as the caterer and client together turn a vision into reality. The science is the business of measuring money, manpower, and material. Successful off-premise caterers recognize the importance of both aspects—art and science—and are able to work at both the creative and the financial levels. In off-premise catering, there is only one chance to get it right. Many events, such as wedding receptions, occur only once in a lifetime. Other events are scheduled annually, quarterly, or on a regular basis, and the caterer who fails to execute all details of such an event to the satisfaction of the client will seldom have another chance.

Unfortunately for some, off-premise catering can be like living on the brink of disaster unless they are experienced. Uninitiated amateurs may not recognize a volatile situation until it becomes a problem, later realizing they should have recognized it earlier. Catering off-premise is very similar to a sports team playing all of its games away from home, in unfamiliar surroundings, with none of the comforts of home to ease the way. There is no home field advantage, but there is a minefield disadvantage! As caterers plod their way toward the completion of a catered event, there are thousands of potential "land mines" that can ruin an otherwise successful affair.

Off-Premise and On-Premise Catering

There are some difference between off-premise catering and on-premise catering. These differences are explained below.

Client's Viewpoint

Most clients fail to consider the cost of the rental equipment such as tables, chairs, linens, china, glassware, and flatware when they consider engaging an off-premise caterer. They think it will be less expensive to entertain in their homes, or at unique off-premise sites, than in hotels. In fact, it can be more expensive, considering not only the cost of the rental equipment, but also other costs such as transportation of food and supplies to the site, the costs of special labor and décor, the need for tenting, air-conditioning and/or heating, and other expenses.

Clients may save some money by buying their own liquor, but this can be insignificant as compared with the added costs. For many clients, the additional costs are far outweighed by the benefits of entertaining in the privacy of their own homes or the uniqueness of a special off-premise location such as a museum, state-of-the-art aquarium, antique car dealership, or historical site.

Caterer's Viewpoint

Off-premise caterers must plan menus that can be prepared successfully at the client's location. For example, foods to be fried should not be cooked in unventilated spaces, like small kitchens in high-rise office buildings. On-premise caterers are not as limited in this regard, and they are generally supported by built-in equipment that can support a wider variety of menus. On-premise party personnel are more familiar with the party facilities than those who work at a variety of unfamiliar locations.

Off-premise catering generally has greater seasonal and day-to-day swings in personnel needs, which can create a greater challenge for the off-premise caterer, who is constantly recruiting and training staff; turnover is usually high because such work is on an "as-needed basis." There is definitely a greater potential for oversights in off-premise catering. Backup supplies, food, and equipment can be miles away or even inaccessible when catering, for instance, aboard a yacht miles from shore. In spite of the uncertainties, off-premise catering offers the opportunity to work in a greater variety of interesting locations. The work is more likely to be different each day, resulting in less boredom and more excitement. For those looking for unlimited challenges and rewards, off-premise catering may be the answer.

Advantages and Disadvantages

There are numerous advantages of catering:

— Advance deposits

— Limited start-up investment

— Limited inventories

— Controllable costs

— Additional revenues

— Business by contract

— Direct payment

— Advance forecasting

— Free word-of-mouth advertising

— Selectivity

Each of this will be described below.

Most off-premise caterers require some form of advance deposit prior to an event. This deposit provides the caterer with some security if the event is canceled and also can be used to purchase some or all of the food and supplies for the party. There is no need for large amounts of capital to get started, since most off-premise catering operations begin by using the existing kitchen facilities of a restaurant, club, hotel, church, or other licensed foodservice business. In addition, all of the necessary catering foodservice equipment such as china, glassware, flatware, tables, chairs, and linens can usually be rented, thus avoiding having to invest in expensive equipment inventories.

Food and supply inventories, as well as operating costs, are much more easily controlled, because clients must advise the caterer in advance as to the number of guests that are expected. Off-premise caterers need buy only the amounts necessary to serve the event, unlike a restaurant where there is a large variation from day to day regarding the number of patrons and their menu selections. Off-premise catering generates additional revenues for existing operations like hotels, clubs, and restaurants. They can generate even more profit by providing other services—rental equipment, flowers, décor, music, entertainment, and other accessory services.

Both the client and the caterer have expectations regarding the outcome of the party. These expectations should be clearly spelled out in a written contract. Payment for an event is normally made directly to a manager or owner, eliminating a middleman, whether it's a wedding planner, on-site food and beverage director, or one of the caterer's own staff members. This form of direct payment provides for better cash control and fewer folks to share the profit. Advance forecasting is more accurate for off-premise caterers, because parties are generally booked weeks, months, or years in advance. Moreover, each part of the country has seasonal swings, which make revenue forecasting somewhat easier.

Off-premise events generate tremendous amounts of free word-of-mouth advertising, which can produce future business without the necessity of advertising. Many off-premise caterers feel that satisfied guests at one party will either directly or indirectly book another party by speaking favorably to friends and co-workers about the event and the caterer. In other words, one party can create future parties. Caterers also have the advantage of being somewhat selective about their clients. There are no laws that require you to accept every request to cater. If the job doesn't meet your standards, politely decline.

In sticky situations where you've already begun to work with a client but find that your communication styles just don't mesh—or, as sometimes happens with weddings, the client is not heeding your advice and you can't even decide who's really in charge—you can walk away, as long as you do so within the terms of your written agreement. Off-premise catering does have some disadvantages too: Catering managers, owners, and staff undergo periods of high stress during very busy periods. Deadlines must be met. There are no excuses for missing a catering deadline. Stress is compounded because the workload is not evenly spread throughout the year.

For most off-premise caterers, 80 percent of the events are scheduled in 20 percent of the time. For most, weekends are generally busier than weekdays. Certain seasons, including Christmas, are normally busier than others. Of course, caterers must maintain general business hours too! Many have left the catering field, burned out by the constant stress and high energy demands. The seasonality of the business makes it difficult to find staff at certain times. Revenues are inconsistent, making cash management very difficult, particularly during the slower periods when expenses continue yet revenues do not. For those caterers who operate hotels, restaurants, clubs, and other businesses, the time away from the main business—spent on the off-premise business—can hurt. It is difficult for even the most well-organized person to be in two places at the same time.

Many hoteliers and restaurateurs find the rigors of off-premise catering too great. Some quit after realizing the difficulty of catering away from their operations. They feel that the financial benefits are insufficient compared with the effort required to cater off-premise events.

Factors Affecting the Success of Off-Premise Catering

Work experience

Prior experience in the catering profession or the foodservice industry is important. Experience in food preparation and foodservice helps caterers understand the procedures and problems in both areas and how the two areas interface. Those with a strong kitchen background, for example, would be wise to gain some front-of-house experience, and front-of-house personnel should learn the kitchen routine. Many successful off-premise caterers began by working as accommodators. Accommodators are private chefs who are hired to prepare food for parties. Many assist the client with planning the menu, purchasing the food, and even arranging for kitchen and service staff. The food is prepared and served in the client's home or facility, eliminating the need for a catering commissary. Accommodators receive a fee for their services. The party staff is paid directly by the client.

Passion

Successful professionals are passionate about their work, and caterers are no exception. They love what they do! Clients and staff members will quickly detect a lack of passion, and it will cost you business and good workers. If you don't love what you do, move on and try something else.

An entrepreneurial nature

The desire to be an entrepreneur is a trait that is highly desirable for off-premise caterers. An entrepreneur must be willing to spend extraordinary amounts of time and energy to make the off-premise catering business successful, possess an inherent sense of what is right for the business, have the ability to view all aspects of the business at once rather than focusing only on one or two parts, and demonstrate a strong, incessant desire to be his or her own boss and become financially independent.

Business Knowledge

— Accounting and bookkeeping skills are necessary to understand the financial aspects of operating a catering business. The ability to prepare and interpret financial statements is essential.

— Learn as much about computers as you can. You'll be amazed at how much you can accomplish by using e-mail, having a website, and using specialized programs for everything from budgeting to menu planning.

— It's also important to understand the legal aspects of catering. Laws that affect caterers include regulation of licensing, contracts, liability, labor, and alcoholic beverage service.

— A caterer, like any other businessperson, must have some human resource skills. Knowing how to recruit, train, motivate, and manage personnel is critical.

— Off-premise caterers should be knowledgeable about how to develop and implement a marketing plan.

Ability to plan, organize, execute and control

These are the four basic functions of management. To plan, a caterer must visualize in advance all of the aspects of a catered event and document the plans so they are readily understood by the client and easily executed by the staff. Organizing is simply breaking down the party plans into groups of functions that can be executed in an efficient manner. Execution is the implementation of the organized plans by the party staff. Controlling is the supervisory aspect of the event. All well-organized and well-executed plans require control and supervision. The adage is, "It is not what you expect, but what you inspect." The premier off-premise catering firms in the United States insist on excellent supervision at each event.

Communication abilities

Listening is the key to good communication with clients and prospective clients. Off-premise caterers must listen carefully and attentively to determine what the client needs. Astute caterers must be able to respond to client requests in such a manner that the client will immediately gain confidence in the caterer. Communicating with staff is a complex issue. In simple terms, it can be reduced to the ability to tell staff what is expected so that they understand, and the ability to receive their feedback regarding problems, both actual and potential. The result of effective communication is an off-premise catering staff that professionally executes a well-planned party that meets or exceeds the client's expectations.

Willingness to take risks

Off-premise catering is a very risky business. It is not for the fainthearted who are afraid of the unknown. For example, it is more risky catering a corporate fund-raiser at the local zoo under a tent than serving the same group in a hotel ballroom. Off-premise caterers must know when the risk outweighs the gain. In this particular example, catering the event at the zoo without adequate cover in case of rain would probably be too risky. The event could be ruined. The tent makes the risk of rain a calculated one.

Working without rest

Off-premise catering requires working long hours without rest or sleep, lifting and moving heavy objects, intense pressure as deadlines near, and even long periods of little or no business, which can cause concern. Successful caterers should be in good physical shape, have a high energy level, and be able to mentally deal with seasonal business cycles that range from nonstop activity to slow periods with little or no business. Off-premise caterers must be self-confident, but at the same time realize that they must always find ways to improve the quality of their food and services. In this profession a fondness for people and feeling comfortable in crowds is important. A "cool head" when under pressure will keep both staff and client calm while potential problems are resolved professionally and efficiently.

Creative caterers

This is the benchmark of all outstanding caterers. Creative caterers are able to turn a client's vision into reality by creating the appropriate look, feel, menu, service, and ambiance. Those who are not very creative can learn to be, or they can employ those who are creative.

Dependability

Dependability is a major cornerstone of success in off-premise catering. When a caterer fails to deliver what was promised, the negative word of mouth travels fast among clients and potential clients. Even in those situations where circumstances change, making it more difficult to perform as promised, the outstanding caterer will find a way to deliver rather than use the changed circumstances as an excuse not to deliver.

Open-mindedness

Open-minded caterers read up on catering trends and try new recipes and menus. They are willing to prepare unfamiliar dishes requested by clients, after thoroughly testing and understanding the recipes. They discover and try new dishes. They are always learning better ways to run their businesses.

Ability to manage clients' needs

The needs of the client must always come first. Success in this business comes from identifying these needs and satisfying them. Unsuccessful off-premise caterers are those who get lost in trying to satisfy their own needs for money, equipment, and greater self-esteem. They forget that the primary goal is to serve the needs of the client. When a client's needs are met, the caterer's needs for revenues, profits, and positive feedback will automatically be met.

Ability to project a favorable image

Prospective clients hire caterers based on their perceived image of the caterer and what the caterer will provide. In some sense, then, caterers are selling themselves more than their food. Off-premise caterers must be able to project a favorable image to the client, one that is in accord with the client's expectations. For example, a caterer whose image is sophisticated and upscale will be hard-pressed to sell a Little League banquet with a low budget. Successful caterers understand their projected images and target their marketing efforts at those clients who desire that image.

Sense of humor

In this pressure-packed, deadline-oriented, and stressful business, it is easy to get carried away with the magnitude of the undertakings and become so tense and uptight that work ceases to be fun. Laughter at the right time can relieve that tension and stress, putting a renewed sense of fun into the work at hand.

Management of Off-Premise Catering

Even those who possess the qualities that indicate off-premise catering success must know how to put these talents to use effectively. Off-premise caterers should be hands-on managers who are constantly customer focused. They must be able to lead staff and clients alike, while conducting business in a professional manner. They must be able to make timely, ethical decisions, while understanding what makes for a successful event. They must also avoid those situations that cause a business to fail.

That's the reason you need a strategic plan—a roadmap to help you determine the direction in which you wish to go, and the specific goals you'll need to accomplish to get there. A strategic plan starts with a statement of core values, which may include things like client satisfaction; ethical business practices; staff satisfaction, training, and motivation; community service; and operating an environmentally conscious business.

From these core values, a caterer can develop a Mission Statement—a succinct sentence that sums up the company's mission. Here's an example: "To meet the catering needs of the corporate community, providing high levels of service and food quality that result in repeat business and vital growth." After the Mission Statement comes the Vision Statement—a concise summary of where you want to be in the future.

Your Mission and Vision Statements lead naturally to the next step—to establish goals for the operation. You may have heard time management experts use the term "SMART" when describing goals. The acronym stands for:

Specific: The goals to be accomplished must be easily understood, concise, and unambiguous.

Measurable: There should be no question about whether one attains, or falls short of, a goal. It may be measured in terms of quality, cost, quantity, or time. Attainable: The goals may be just out of reach, but they're not out of sight! The best goal challenges and motivates you and your team. If it's practically impossible, it may be too frustrating.

Relevant: The goals must fit well with your long-term mission and vision, your objectives, and the results you expect.

Time-bound: There must be a specific deadline for completion of each goal. An example of a SMART goal might be to increase sales and profits by 20 percent each year for the next five years.

Once a caterer has set goals, there must be certain trade-offs. To increase sales, for instance, may require raising prices, hiring more staff to be able to cater more events, or spending money on advertising. The major goals can be broken into smaller, intermediate steps, with a time line to keep the company on track. And remember, goals are not just for the owner of a company. The staff and other professionals employed by the company—tax preparer, banker, attorney—should also be well aware of the goals. You'll need their help to achieve them, and you want them on your side, committed to your goals.

Too often, caterers believe they can do everything themselves. They fail to ask for or accept advice from outside consultants and colleagues. It is far more intelligent to ask for assistance when you need it. Someone familiar with your plans and your passion for them is far more likely to be helpful. Finally, as soon as a goal is set, take some action on it. The last part of a strategic management process is to reevaluate your mission, vision, and goals periodically. Times change, trends change, and you become aware of new information. Let's say a caterer's sales year showed a 50 percent increase, when he or she had set a 20 percent annual goal. In this case, the next year's goal might be more realistically revised to a 30 percent increase.

In catering, the details are virtually endless, a stream of tiny elements that might go wrong and result in a catastrophe. One thing forgotten, misheard, or misplaced can ruin an event. So it's important to check and recheck and to be prepared for last-minute emergencies. It is simply not possible to run this kind of business from behind a desk, reading computer printouts and delegating all tasks. Off-premise catering companies must be managed from the center of the action, whether that is with the guests or preparing foods in the kitchen. It comes from checking and rechecking every detail to ensure that it meets the highest of standards. It comes from inspecting for the best and expecting the best. Some call this management style "management by walking around." In one sense that is true, but there is more to it than walking around. Astute off-premise caterers must:

— Obtain feedback from clients and guests regarding the food and service. Oversee the catering staff to ensure they are performing as directed and as expected.

— Help out when a table needs to be cleared or when the bar suddenly becomes very busy. Help in the kitchen during critical times such as hot food dish-up, and even help scrape, stack, and wash dirty dishes if that's what is necessary.

It's a roll-up-your-sleeves kind of profession, and you should never be totally satisfied with the way things are. Always look for new ways to present food and make it more flavorful, and for better and more efficient ways to do things.

Customer-Focused Management

An off-premise caterer's full-time mission must be to satisfy the needs of clients.

Decision Making

Off-premise catering managers must make decisions that keep their operations running smoothly. They realize that some decisions will be better than others, that there is no perfect solution to every problem, and that the best decision-making goal is to find the best possible solution with the least number of drawbacks. Connie Sitterly, a management consultant and author, states that to be a good decision maker you should "plan ahead so when problems crop up, you're prepared to act, not react. Control circumstances, instead of allowing them to control you. Take the initiative by anticipating and solving business problems."

— Remember that there's seldom only one acceptable solution to the problem. Choose the best alternative.

— Make decisions that help achieve the company objectives.

— You need to consider feelings whenever people are involved. Even if you must make an unpopular decision, you can minimize repercussions if workers know you have taken their feelings into account.

— Allow quality time for planning and decision making pick a time when you are energetic and your mind is fresh.

— Realize that you'll never please everyone. Few decisions meet with unanimous approval... the appointed authority, not the majority, rules.

— Make time for making decisions in business, delaying a decision can cost thousands of dollars.

— Put decision making in perspective. Every executive feels overwhelmed at times by either the enormity or the number of decisions made during a business day.... For peace of mind accept that you are doing the best job you can with the time, talent, and resources you have.

— Don't wait for a popular vote. Rallying your colleagues around your decision before you take action or waiting for their vote of confidence before deciding anything may cost too much in time. There are times when you just have to do something.

Leadership Skills

There are major differences between those who lead and those who manage. Chattering companies need both types of executives, and some who can do both. If a catering company is earning seven- and eight-figure annual revenues, it is most definitely being led by people with leadership skills. Leaders are able to get people to do things they don't necessarily like to do, but they do them and even enjoy them. We can say that;

A Manager	*A Leader*
Maintains	Develops
Administers	Innovates
Relies on systems	Relies on people
Counts on controls	Counts on trust
Does things right	Does the right things
Works within the system	Works on the system
Manages things	Leads people

A leader is more like a thermostat than a thermometer. A thermostat sets the standard temperature for the space it's in. A thermometer simply records the temperature; it can't change anything. And one more important trait: Leaders take a little more than their share of the blame and a little less than their share of the credit.

Professionalism

Off-premise caterers who are not professional in their business practices will never reach the pinnacle of success in the field.

Ethics in Management

The Roman philosopher Publilius Syrus said, "A good reputation is more valuable than money." This is as true today as it was in ancient times. And yet, lack of ethics is perhaps the most widely discussed topic in today's business world. We read and hear of illegalities, scandals, and other forms of questionable behavior bringing down some of the nation's largest corporations. Off-premise caterers are in no way exempt from ethical concerns. Even the smallest caterers deal in issues of fairness, legal re- quirements, and honesty on a daily basis. Examples include truth in menu, misleading advertising, unexpected and unjustified last-minute add-ons to the party price, and even underbidding a competitor when the client has disclosed your competitor's price.

The truly ethical caterer will assume responsibility for the host to ensure that the host plans an event in the best interest of the guests. A host who wishes to serve alcohol to underage guests or barbecued ribs to a group of elderly people (tough to eat with dentures) is out of line and needs to be advised that this will not work. In fact, an ethical caterer will refuse to cater an event that is clearly not being planned in the best interest of the host or guests. There are times when a caterer is given a free hand in planning a menu. Perhaps a grieving client calls for food after the funeral of a loved one, saying, "Please send over food for 50 guests tomorrow night. You know what we like!" The ethical caterer will not take advantage of this situation by either providing too much food or overcharging the client.

Another temptation arises when the caterer is pressed to cater more events on a certain day or evening than he or she can reasonably accommodate. The extra money looks good. Unethical caterers will rationalize that they can handle all the events, even if an inexperienced supervisor or staff must oversee these events, or even if the kitchen staff will not be able to prepare the caterer's usual high-quality food because of lack of time and personnel. Caterers who take on more work than they can reasonably accommodate are greedy and are considered by many observers to be unethical.

In the foregoing situation the caterer should decline the work and perhaps recommend another caterer. Some caterers refuse to recommend another catering firm because they feel that if the client is not pleased with the other firm, the caterer who turned down the business will be blamed for the recommendation. Other caterers freely recommend one or more companies when unable to cater events. There are times when it is very hard not to bad-mouth a competitor, but this is considered unethical as well as rude. Those who are ethical would rather point out their own strengths than downgrade the competition. It can be very tempting for self-employed caterers to underreport income or overstate expenses. They rationalize that no one will know if they accept cash for a party, then fail to report it as income and pay the associated tax, or that no one will know if they happen to charge personal expenses now and then to the business.

Some caterers who are licensed to sell liquor by the drink or by the bottle are tempted to bill clients for beverages that were not consumed. These practices are not only unethical—they are illegal. Other ethical violations occur when caterers receive under-the-table cash "kickbacks" from suppliers, misrepresent their services to potential clients, or bid on party plans or ideas stolen from other caterers.

Caterers also soon learn that some clients are unethical. A few are masterful at finding fault with a wedding or other important event, then demanding a "discount" based on whatever flaw they feel they have uncovered. Some will refuse to pay for linens that were damaged by candles they lit on them! You'll find people who, midparty, will ask you to stay "a couple hours of overtime, just to wrap things up"—then not show up to pay you for the extra time the next day, as agreed. Others will haggle over the tiniest details on an invoice or try to engage more than one caterer in a bidding war to lower prices. Caterers who deal with "middleman" organizations, like destination management firms or production companies, may find that a client of one of these companies will come back later to try to deal directly with you, thus cutting out the middleman who recommended you!

As a catering professional, you need to expect a certain amount of this behavior and must protect yourself if you suspect an ethical question may arise. Insisting on security deposits, having a valid and authorized credit card number on file for unforeseen charges, refusing to look at other caterers' written bids, and standing firm on your own invoice prices are just a few ways ethical problems can be avoided. And rather than cut out a legitimate middleman-type of vendor, you can either refuse to deal directly with a client who tries such a maneuver or suggest a commission be paid to the middleman.

You will also be put in some sticky situations as—during tough times, and even good times—certain clients will make unrealistic requests. They've often been good, regular clients too! But they'll promise you future business if you'll cater their party "at cost," or defer payment for them, or ask some other special favor "just this once." These requests are unfair, and you're right to be squeamish about them. Off-premise caterers should be extremely wary when approached in this fashion. As a general rule, clients who do not pay their bills in a professional manner, or who are not willing to pay a fair price for catering services, are not worth the headaches they cause.

The Jefferson Center of Character Education has set forth a list of ten "universal values": honesty, integrity, promise keeping, fidelity, fairness, caring for others, respect for others, responsible citizenship, pursuit of excellence, and accountability. These values should provide some solid guidance for any businessperson who considers him- or herself a true professional.

Great caterers do more than imitate—they innovate. There are distinct advantages for those who offer a unique menu, a unique service, or perhaps a unique location. They may build and improve on someone else's concept, but they strive to take the idea to the next level. Rather than mimicking another's success, they imprint their own signature on their menus.

The "Unique" bar may include all the traditional accompaniments too—but what a difference a little imagination makes! There might even be a bit of caviar to top the mashers at the Unique bar, and perhaps they'll be served in martini glasses.

As with any career, catering professionals need to reexamine their business strategies from time to time. Some caterers do what they do best, are well known for it, and never vary their formulas. Their clients love them and get exactly what they expect. Other caterers blindly copy everybody else. They ricochet from one recipe to the other, never bothering to see if it meets their clients' needs. If they read about it in Food Arts magazine, they feel they have to serve it! But most caterers lie somewhere between these two extremes, blending the successful ideas of the past with new twists.

Great caterers also separate themselves from competitors by using the resources around them to build their businesses. In South Florida, for example, one caterer specializes in event planning for doctors, through his hospital foodservice management job. Another has an exclusive off-premise contract for a sports facility; a third was the on-premise caterer for a city club, which resulted in off-premise jobs for the club members.

Personal Management

Off-premise caterers must learn how to deal with principles of stress management, time management, and personal organization if they are to manage at peak efficiency. Time is our most precious commodity, and to waste it because of being overstressed or disorganized will inevitably result in less-than-desirable results.

Stress management

Stress comes from interaction with others, and from having to meet deadlines. A certain amount of stress and tension is necessary to achieve the best results—those who are too laid back generally do not maximize their potential—but too much stress causes chronic fatigue, irritability, cynicism, hostility, inflexibility, and difficulty in thinking clearly. Catering managers who are overstressed are unable to perform at maximum capability. Stress can often be controlled through:

— Daily exercise such as brisk walking, running, or other aerobic pursuits that increase the pulse rate. Some folks purposefully take their minds off work when they exercise; for others, the daily walk or run is a time to get their day mentally organized.

— Relaxation techniques, including meditation and yoga.

— Writing down the issues that cause stress. Identify those issues in your life that can be controlled, and simply decide to make the best of those that cannot. List ways to deal with the controllable stress factors.

— Reading articles and books on stress reduction.

It is important to remember that some stress in catering is good. An arrow would not be propelled from a bow if the bow was not stressed. However, too much stress can break the bow, as well as ruin catered events.

Time management

There are only 168 hours in each week, and the greatest rewards come to those who accomplish the most meaningful things during this fixed amount of time. Off-premise caterers realize that if they can accomplish more meaningful production in less time, they will have more time for things other than work. They also realize that working smarter, not harder, through the effective use of time will produce greater results. The key to effective time management is to set goals for a lifetime, for five years, and for each year, month, week, and day. Without written goals, off-premise caterers cannot effectively manage their time. Because time management involves choosing how to spend time, it is impossible to make proper choices without knowing your desired goals. The captain of a ship without a destination cannot choose the proper course. He will cruise aimlessly at sea, never reaching his port of call.

Off-premise caterers can choose from an array of timesaving techniques and technical advances to help them in the quest to efficiently manage time:

— Make those daily, detailed lists of goals and objectives.

— Use technical advances to speed up paper handling, such as fax machines and computers with word processing, accounting, and menu-planning software.

— For heaven's sake, if you don't have a computer, get one! You can purchase one nowadays for a monthly payment of less than $40. You can take classes to learn how to use it or hire someone to teach you individually.

— Use cellular phones to stay in touch while away from the office. These are lifesavers at off-premise catering locations when emergency and other calls are necessary, and if you have downtime, a cellular phone can make it easy for you to use this time to return phone calls.

— Handle incoming papers only once. Here's the rule: Do it, delegate it, discard it, or file it.

— Do your most important work at times when you happen to be most alert. Take advantage of your peak energy periods to handle your most challenging tasks.

— Sign up for a seminar or course in time management to learn more tips.

One of the biggest time wasters for a caterer is also the source of much business—the prospective client who calls to ask questions—so it's an interruption that cannot be ignored, but can be controlled. Whoever answers the phone at your business should always qualify the incoming call by asking:

— The date of the event

— The location of the event

— The number of guests

— The budget for the event

First of all, time can be wasted talking about an event before you ask the date and discover you're not able to do it in the first place because of a scheduling conflict. Perhaps the number of guests is too small or too large for your particular company, the budget is insufficient, or the proposed location is already booked for another event.

Always focus on results by asking yourself, "Will this activity help me achieve any of my goals?" Prioritize tasks in order of their importance and know when to delegate them to others. Most people waste countless hours, days, weeks, and years chitchatting on the phone, shuffling papers, running errands, and doing other things that are easy enough but offer little or no payoff. Learn to delegate these types of tasks whenever possible. Pay other people to do them, and don't tell yourself you can't afford it—you can always make more money, but you have only so much time.

The true achievers—in catering and in other fields—minimize their time on low-priority, low-payoff tasks and turn their attention to those things that will bring the greatest rewards. These tasks are often difficult to accomplish, take a great deal of time, and involve at least some risk. For example, a caterer could spend the entire day showing prospective clients numerous suitable locations for a major event. The caterer would then spend the next three days preparing a written proposal for an event at each of the locations, with no guarantee that the event will even take place. However, if the caterer is hired, there's a five-figure profit to be made. Worth the risk? Certainly! Another high-payoff task might be to write a new catering menu. Both this and the aforementioned task require large chunks of time and involve some risk, but more than likely will produce major rewards in increased revenues and profits.

When projects, tasks, catering kitchens, and offices are organized, things run much more smoothly and efficiently. The time spent looking for things and jumping from job to job is wasted time that could be put to much better use. Many off-premise caterers have found various methods that work for them:

— Establish a filing system using hanging folders and manila folders. Categories can include upcoming events, projects to do, and projects pending. Files should be stored vertically, rather than stacked atop one another, for greater accessibility.

— Take a tip from event planners who start a separate notebook for each event they are working on. Into this three-ring binder go all notes, contracts, sketches, color samples—anything for that particular job.

— Consider hiring a professional organizer to come to your office and set up a filing and record-keeping system that works for your business.

— Keep those items that are used frequently close by.

— Focus on one project at a time, rather than jumping from one thing to another. This can be easily accomplished by blocking out some time during the day to work on major projects and arranging for no interruptions.

— Whenever possible, try to schedule time to return phone calls and/or e-mail messages. That way, you can handle them all at once, instead of scattering them (and your thoughts) in five-minute intervals throughout the day.

— Either at the end of each day or first thing in the morning, prepare a list of things to do for the day.

Catering in the Future

Catering is neither rocket science nor brain surgery. Change is inevitable in this business, but not at the same rate as, say, in molecular theory or medical technology. In fact, in catering, rediscovering foods of the previous century is trendy! Many caterers still feature the signature dishes—honey coconut shrimp, beef tenderloin, Caesar salad—that they've served for decades. Why? The customers demand, and enjoy, them. This certainly doesn't mean things stay stagnant in our industry. Innovative buffet and food station décor will continue to evolve. Most catering companies will continue to build their reputations on elegant, "over-the-top" food presentations, and the healthy competition shows no signs of abating. Other caterers prize research, developing cutting-edge menu items to set them apart from the pack.

More women are entering the off-premise catering field. Staffing woes will continue to be monumental, as hiring, training, and retraining get tougher. Foodservice has always been a somewhat transient industry. Astute caterers will use pre-employment aptitude and personality testing, master online staff scheduling systems, and develop their own training programs. They will also realize, if they haven't already, that they must treat their employees at least as well as they treat their clients. Along the same lines, in a top-tier catering operation, the employees treat each other as well as they treat their clients.

Caterers of the future will come to realize that bigger is not necessarily better. Having a large volume of business is admirable—but only when the quality of your work rises to the same level. A company can grow to the point where quality slips, gross profit margins lag, more equipment is needed, overhead costs expand, and the bottom line shrinks proportionately. The intelligent caterer will downsize, watch margins and profits grow—and overall stress levels diminish—as they become more selective about the clientele they service.

Caterers are realizing that "high tech" will never replace personalized service, or "high touch"—but without high tech, they'll limit their potential for high touch. In an industry where, amazingly, some caterers still don't accept credit cards, the savvy businessperson is learning to embrace new technology, launching interactive websites and e-mail marketing campaigns. They're creating improved computer-generated proposals, rental orders, packing lists, staffing schedules, and instant financial statements. And they're realizing that computer-savvy business owners have more time to do what they love—which is run their business! Competition will continue to increase. Sales will grow, but not without some dips, because economic woes, terrorist attacks, and the resulting fears cannot help but impact the catering profession.

More caterers were hurt financially by the recession at the beginning of this century than by the September 11 terrorist attacks, but both left their marks on the industry. An increased use of security cameras at high-profile events (and in some cases, to thwart theft) is one result of the heightened awareness. Mega-event catering is acknowledged as an excellent way to grow business—at golf and tennis tournaments, NASCAR races, air shows, boat shows, and more. In addition to being profitable events, they expose the caterer to a wider range of potential clients. Then again, a caterer from Augusta, Georgia, generates enough revenue from serving sandwiches and beverages at the Masters' Golf Tournament that he need not cater at all the rest of the year! The pressure experienced in servicing huge, multiday events is as big as the events themselves, but the rewards can be significant.

Habits of Successful Caterers

In this section we can see some additional techniques, philosophies, and real-life ways to be successful in the challenging field of off-premise catering.

Willingness to Take Calculated Risks: One of our favorite sayings is, "A turtle goes nowhere until it sticks its neck out." In order to succeed, we must be continually growing and improving, and the only way to do this is to leave our comfort zones—and stick our necks out! If you're right-handed, you feel quite comfortable writing with your right hand. Try writing with your left hand. You're definitely out of your comfort zone. But after a while, you'll find you can actually write with either hand. Successful caterers make things happen by taking calculated risks, whether it is trying new menu items, new buffet display concepts, or accepting a job in a new and challenging off-premise location. Caterers who refuse to take risks fail to grow and learn are left behind.

Sincere Concern for Others: Nobody cares how much you know until they know how much you care. Empathy and genuine concern for your clients and staff are paramount to long-term success. What are their needs, wishes, and desires? What are their concerns and their "hot buttons"? By putting ourselves in their positions, we can begin to show concern for others and understand them. When we do this, we develop meaningful relationships and, not coincidentally, loyalty. We give them what they want, and we get what we want.

Keeping Up with Current Trends: It's not just a matter of food and presentation and theme trends. Caterers who are not wired to do business online through the Internet and e-mail are missing out on huge opportunities. The online catering referral service, Leading Caterers of America (founded by the book's coauthor Bill Hansen), receives 5 to 20 inquiries per day from clients looking for catering services coast to coast, in Alaska, Hawaii, and occasionally overseas. People do shop for catering online, and the companies that lead the way have high-quality websites and diligently reply to e-mailed requests in a timely manner.

Caterers need to get in the habit of responding to e-mail correspondence as soon as possible, as well as providing e-mailed proposals to those clients who prefer to do business via their computers. Event planners who book caterers for their clients love receiving e-mailed proposals, because they are easy to copy-and-paste into their own proposals. If you're not in the habit of working online, you're behind the times.

Excellent Priorities and Time Management: You get 20 percent of your sales and profits from 80 percent of your clients, and 80 percent of your sales and profits from 20 percent of your clients. None of us ever go home at night thinking that all the work is done—it never is. It's simply a question of what's most important, as well as what's most urgent. Urgent things are never really an issue. There's no question that if you have a catered event today, it will get done. But what's most urgent is not necessarily what's most important. You must understand the difference.

For example, you could spend a day catering three small parties for 25 guests each, but fall behind on preparing a proposal for another job, in three months, for 500 guests—and lose it to a competitor whose proposal was simply submitted on time. Successful caterers spend their time in those areas that generate the biggest pay-backs in terms of money, quality, and other rewards. They make a habit of planning their days, leaving time for the most important, as well as the most urgent. At the start of each day they prepare an agenda that details both short-term objectives and long-term goals. If you're a student, you should already be using this technique to accomplish as much as you can in school.

Quality before Quantity: Bigger is not necessarily better. Still, many of us get caught up in that way of thinking. If our sales are $1 million, let's go for $2 million. If they're $2 million, what's wrong with $4 million? There's nothing wrong with building sales if quality does not suffer. However, when the quality of our products and services suffers so does the quality of our lifestyle. More business means more hours at work. And doctors will tell you they've never met a man or woman who, on a deathbed, expressed a wish that he or she had spent more time at work.

If we can grow our businesses with no adverse effects on the quality of our lives or our products, then we should go for it! But if we find profits slipping and clients complaining, and we need a letter of introduction when we stumble home at 3:00A.M., then something's very wrong. We need to make of habit of continually asking ourselves whether we might be better off with less business and more time for ourselves and for our families. We need to continually examine the quality of our work to ensure that it's not slipping because we've allowed ourselves to take on too much.

Being Detail Oriented: Being aware of the details in flavors, looks, aromas, and tidiness separates the average caterers from the superstars. And, by all means, check the spelling, grammar, and punctuation in all your written materials, from brochures to contracts—or hire someone to do it. Again, the goal is to present a professional image.

Setting High Standards: If you refuse to accept anything but the very best, you very often get the best. Successful caterers set their standards high and expect excellence from themselves and their staff members. They're never happy with the status quo, always striving to make each party, wedding, or event better than the last. They debrief after an event, asking staff for input and improvements. They know that if they fail to improve, they're leaving the door open for their competitors to capture a good customer or a larger share of the market.

Successful caterers also make a habit of lifelong learning. They're forever reading, attending trade shows, and exploring areas that will help them improve their own businesses with new ideas. They challenge and reward their staff members for having the same attitude.

Vince Lombardi, the late NFL coach, who during his career coached the first team to ever win the Super Bowl, put it this way: "The quality of a person's life is in direct proportion to their commitment to excellence, regardless of their chosen field of endeavor."

Evaluation of Off-Premise Caterer Success

There are a number of signs to look for when evaluating an existing off-premise catering business. Healthy companies rate highly in all of these areas. Those that are unhealthy, or even on the brink of failure, will not rate nearly as well.

— Management thoroughly plans, organizes, executes, and controls each catered event.

— Proper controls are in place for costs, accounts receivable and payable, and liquid assets such as cash and inventories. Theft prevention is also a priority.

— Food and service quality is well-controlled and meets or exceeds clients' approval.

— Pricing for food and services is fair and competitive with other firms in the marketplace. There is a spirit of healthy competition.

— The catering firm enjoys good working relationships with both clients and suppliers.

— Time and attention are given to food safety in storage, preparation, and display. Employees know the local health codes and follow them.

— There is sufficient working capital to operate the business. The firm can make loan payments as they become due. Excessive credit is not extended to clients.

— Budgets are prepared and followed. Business records, insurance coverage, and licenses are kept up to date. The information derived from these records is used to provide data to help manage the business.

— Sales growth is controlled. There are sufficient financial and personnel resources to operate as business steadily grows.

— Market trends are anticipated.

— Management and staff have a good working knowledge of the off-premise catering field.

— There are solid, trusting relationships between management and staff. Staff members are well trained and feel truly appreciated—because they are.

— Management works closely with a qualified accountant to plan for payment of taxes.

— And, finally, management is willing to seek qualified professional assistance if problems arise.

Model of an Off-Premise Catering

Figurel is a diagram of all the factors that enter into the off-premise catering arena. It shows how managerial philosophies and techniques, as well as laws regarding personnel, business, alcoholic beverage service, and sanitation and safety, must all be interrelated to guide the company. It then depicts how marketing efforts produce clients, which in turn creates needs for site inspections and logistical plans, including planning in these specific areas: menus, beverages, equipment, personnel, and any other related services.

Once the planning is complete, it is possible to provide clients with written proposals, which include all the aforementioned plans along with pricing. Normally, proposals are modified somewhat. Once modification is complete and all provisions meet with the approval of both caterer and client, a contract is prepared that contains all the conditions outlined in the proposal. As the party date approaches, certain operational elements are addressed, such as:

— Hiring and scheduling staff
— Purchasing and pre-preparation of menu items
— Ordering equipment as needed from party rental companies
— Obtaining licenses and permits, as needed, for use of the site, serving alcohol, etc.
— Preparing a "pull sheet" that details all items supplied by the commissary to produce the party.
— Coordinating all beverage and accessory services with the client and the vendors.

All the preplanning elements culminate on the day or night of "The Show." That's when staff, equipment, food, and other services arrive at the party site, and the event is executed. After the event, there are certain outcomes, which include:

— Positive and/or negative word of mouth about the event
— Revenues, expenses, profits, and cash
— Accounting records

8

Scenario of Beverage Industry

In each country, the bar and beverage business is the most glamorous and profitable businesses. In United States the bar and beverage business has gone from an illegal enterprise, carried on behind the locked doors of a speakeasy, to one of the profitable businesses. Together with the food service or restaurant business, the two form the country's fourth-largest industry. In fact, it is impossible to separate them. In the 1930s, the United States was nearing the end of Prohibition, an unsuccessful attempt to regulate alcohol consumption by outlawing it entirely. Such attempts have never worked, because people find other ways to get what they want. And from earliest times, human beings seem to have wanted alcoholic beverages.

History of Wine

Perhaps 8,000 to 10,000 years ago, someone discovered that fermented fruit or grain or milk or rice tasted good, made one happy, or both. The Bible mentions wine consumption in both Old and New Testaments. With all its benefits and hazards, alcohol was a universal feature of early civilizations. At least one legend claims that wine was discovered accidentally, by a neglected member of a Persian king's harem. She attempted to end her loneliness by ending her life, drinking from a jar marked "Poison." It contained grapes that had fermented. She felt so much better after drinking the liquid that she gave a cup of it to the king, who named it "the delightful poison" and welcomed her back into active harem life.

Early peoples all over the world fermented anything that would ferment-honey, grapes, grains, dates, rice, sugarcane, milk, palms, peppers, berries, sesame seeds, pomegranates. We know that grapes were cultivated as early as 6000 B.C., both in the Middle East and Asia. The Egyptians, Phoenicians, and Chinese were tending their vines at about the same time. It is believed that the ancient Greeks got their viticulture knowledge from the Egyptians, beginning to make wine about 2000 B.C.

The practice of aging wines was first discovered by the Greeks, in cylinders known as amphora. Made of clay, they were remarkably airtight. Fifteen hundred years later, the Romans tried a similar method, but their clay was more porous and didn't work as well. So they began coating their clay vessels with tar on the insides, a process known as pitching. By 1000 B.C., grapevines were found in Sicily and Northern Africa. Within the next 500 years, they reached the Iberian Peninsula, Southern France, and even Southern Russia. Conquering Saracen (Arab) tribes in the Middle Ages brought both winemaking and distillation skills with them.

In fact, the words "alcohol" and "still" are Arabic in origin. As the Roman Empire spread, it brought grapes to Northern Europe, too. After the fall of the Roman Empire, the Catholic Church was the most prominent promoter of viticulture. Monasteries became the vanguards of wine production and knowledge, because wine was needed both in everyday life and in sacramental activities. The Portuguese are credited with shipping the first corked bottles of wine to England, but not until the year 1780.

In many cultures, people associated intoxicating beverages with wisdom. Early Persians discussed all matters of importance twice-once when they were sober and once when they were drunk. Saxons in ancient England opened their council meetings by passing around a large, stone mug of beer. Greeks held their famous symposiums-philosophical discussions-during hours of after-dinner drinking. In fact, the word "symposium" means "drinking together." As the Roman historian Pliny summed it up, "In vino veritas". Alcoholic beverages, often in combination with herbs, have been used for centuries as medicines and tonics.

Indeed, herbs and alcohol were among the few ways of treating or preventing disease until about a century ago. But probably the most important historic use of alcoholic beverages was also the simplest: as food and drink. Bread and ale, or bread and wine, were the staples of any meal for an ordinary person, with the drink considered food. For centuries, these hearty beverages provided up to half the calories needed for a day's heavy labor. And they were considered the only liquids fit to drink, with good reason. Household water was commonly polluted. Milk could cause "milk sickness" (tuberculosis). But beer, ale, and wine were disease-free, tasty, and thirst-quenching, crucial qualities in societies that preserved food with salt, and washed it down a diet of starches.

Religious Rites and Wine

Early beers, ales, and wines were considered gifts from the gods, that is, miracle products with magical powers. People used them universally in religious rites, and they still do. The Israelites of the Old Testament offered libations to Jehovah. Greeks and Romans honoured Bacchus, god of wine Christians used wine in the sacrament of communion. Primitive peoples used fermented beverages in their sacred rites. Victories, weddings, and other sacred and joyous occasions were celebrated with "mellow wine" or endless supplies of ale (the word "bridal" comes from bride + ale).

Camaraderie and fellowship were acknowledged with a "loving cup," passed around the table and shared by all until it was emptied.

Of all alcoholic beverages, wine has the greatest religious connection. In the book Religion and Wine: A Cultural History of Wine-Drinking in the United States, the author, Robert Fuller, traces the development of winemaking from the French Huguenots, Protestants who settled along the East Coast of North America in the 1500s, to the Pilgrims in Plymouth Bay in the 1600s to the Franciscan friars and Jesuit priests who built the early missions in California during the 1700s and 1800s. Whether these early Americans were Baptists, Methodists, or Mormons, they permitted and enjoyed limited wine consumption as part of their worship.

According to Fuller, the United States did not have "grape-juice Protestants" until the late eighteenth century. Interestingly, this alternative was first developed by Thomas Welch, a dentist and Methodist minister. His name later became a famous trademark for juice products. At that time, attitudes about alcohol changed as some religious groups began to espouse the theory that the road to heaven required total self-discipline, including the denial of all earthly pleasures. It was feared by some that consuming alcohol would weaken sensibility, ethics, and moral values, and diminish selfcontrol in an age where many churches sought greater control over their members.

History of Beer

As far back as 6000 B.C., Babylonian scribes included beer recipes in their writings, and the ancient Egyptians made note of Ramses III, the pharaoh whose annual sacrifice of about 30,000 gallons of beer appeased "thirsty gods." Before 1850, the beverage preference in the United States was ale, which had been popular in England. It was made like beer, but fermented more quickly, at higher temperatures than beer. The word "beer" comes from the ancient Latin word "biber," a slang term for the beverage made by fermenting grain, with hops for flavoring. Back then, biber was considered lower-class compared to ale, which was made in similar fashion but without the addition of hops.

Just about every civilization has made some type of beer, from whatever grain or root or plant was available in abundance. African tribes made their beer from millet; in Japan, it was rice; in Europe, North and South America, the chief ingredient was barley. The brew was hearty, filling, and provided calories and nutrients to fuel manual labor. The significance of beer in the average person's diet was demonstrated at the landing of the May flower at Plymouth, in what is now Massachusetts. The Pilgrims were headed for Virginia, but the ship was running out of beer. So they were "hasted ashore and made to drink water that the seamen might have more beer," wrote Governor Bradford later.

Beer production and sales played colourful parts of U.S. history. The first American brewery was opened in Lower Manhattan by the Dutch West Indies Company in 1632. There is speculation that the crude streets of New Amsterdam (later, New York City) were first paved to help the horse-drawn beer wagons make better progress, since they were so often stuck in the mud! Brewing became an aristocratic and popular business. William Penn, the Quaker leader who founded the state of Pennsylvania, Revolutionary War leaders Samuel Adams and Ethan Allen, and even George Washington, all were brewery owners. By the mid-nineteenth century, brewing dynasties, that are still household names among today's beer-drinkers had begun in the United States.

Five years later, Frederick Miller purchased an existing facility, Best's Brewery in Milwaukee, Wisconsin. In St. Louis, Eberhard Anheuser purchased a struggling brewery in 1860. His daughter married Adolphus Busch, a German immigrant whose family supplied grains and hops, and the mighty Anheuser-Busch Company was born. A dozen years later, another German immigrant from the Rhineland, Adolph Coors, started to brew beer in Colorado. The Germans brought with them a different brewing style that produced a lighter beer known as lager, which is more pale and clear in appearance than ale and has a drier flavour. Its name comes from a German word for "storage" or "storehouse," since it was routinely stored for several months in cold temperatures before serving.

Making lager-style beer required ice, so it was typically brewed in winter and stored until summer, when the demand was highest. Milwaukee emerged as the nation's brewing center for the most practical reason: ice was easily available from Lake Michigan, and there were plenty of local caves to store the beer. When refrigerators and icemakers were invented, lager could be brewed any time, anyplace. Heat was just as important as cold to the fast-growing beer-making industry. The French chemist Louis Pasteur discovered in the 1800s that, like milk or cider, beer could be heated to sufficient temperature to kill harmful bacteria without diminishing the quality of the brew. This process of pasteurization allowed beer to be bottled for shipment.

Art of Distillation

The art of distillation-first heating, then cooling and condensing liquids to extract and concentrate their alcohol content-was known in crude form even in ancient times. The Chinese and peoples of the East Indies distilled liquids and used the resulting potions for medicinal purposes as early as 800 B.C. About the time the Pilgrims ran out of beer at Plymouth Rock, these forms of concentrated alcohol were coming into favour in Europe. Distilled spirits made from fermented liquids were many times more potent than the original liquids.

The first ones were called aqua vitae (water of life) and were used as medicines, but they were quickly assimilated into society as beverages. Highland Scots and Irish distillers made whiskey. The French distilled wine to make brandy. A Dutch doctor's experiments produced gin, alcohol flavoured with the juniper berry. In Russia and Poland, the distilled spirit was vodka.

In the West Indies, rum was made from sugarcane, while in Mexico, Spaniards distilled the Indians' native drink to make mescal, the great-granddaddy of today's tequila. With increasing supplies of spirits and their high alcohol content, excessive drinking became a national problem in several European countries. In England, cheap gin became the drink of the poor. They could-and did-get "drunk for a penny, dead drunk for twopence," as one gin mill advertised. This particular mill, in the same advertisement, mentioned that it also provided "free straw" (a bed of hay) for sleeping it off. Across the Atlantic, Americans welcomed the new spirits, and it wasn't long before rum became the most popular drink and New England became a leading manufacturer.

George Washington put rum to political use when he ran for the Virginia legislature, giving each voter a barrel of rum, beer, wine, or hard cider. By the end of the century, whiskey was challenging rum in popularity. Washington was once again a forerunner, making his own rye from his own grain in his own stills. However, in 1791, the new U.S. Congress enacted the first tax on whiskey production. Many of the distillers, still trying to recover financially from the Revolutionary War, did not have much money and refused to pay the taxes.

It ended without a shot being fired, but many angry distillers packed up and moved further west to enjoy greater freedom and avoid future confrontations. In fact, relocation to the limestone soils of Tennessee and Kentucky led some of these early distillers to a real gold mine: the cold, clear water supplies of these areas, which are still famous for their whiskey production. The spirit soon became known as "bourbon," since some of the first distillers set up shop in Bourbon County, Kentucky. As the American West was settled, whiskey-easier to store and transport than beer or wine, and much in demand-became a very popular commodity in the trade-and barter commerce of frontier life.

Distillation gained momentum as the process was refined. Rectification, or distilling a liquid more than once, yielded a much cleaner and almost 100 percent pure spirits than previous efforts. Before rectification was perfected, spirits contained flavour impurities. Herbs, honey and/or flowers were added to mask them. After rectification, these items were also routinely added, but now, to enhance the flavour. Some of today's grand liqueurs are the results of these early flavour concoctions. Cognac, for instance, was a pale, acidic French wine for which there was little public demand... until it was concentrated in the 1600s as an eau de vie, French for aqua vitae. It became enormously popular, and still is today.

Pleasures and Politics

Pouring for profit developed hand in hand with civilization. The clay tablets of the Old Babylonian King Hammurabi refer to alehouses and high-priced, watered-down beer. A papyrus from ancient Egypt warns, "Do not get drunk in the taverns... for fear that people repeat words which may have gone out of your mouth without you being aware of having uttered them." Greek and Roman cities had taverns that served food as well as drink; excavations in Pompeii have uncovered the remains of 118 bars. In both Greece and Rome, some taverns offered lodging for the night, or gambling and other amusements.

After the fall of the Roman Empire, life in most of Europe became much more primitive. When next the taverns reappeared, they were alehouses along the trade routes, with a stable for the horses, a place to sleep, and sometimes a meal. In England the public house, or pub, developed during Saxon times as a place where people gathered for fellowship and pleasure. An evergreen bush on a pole outside meant ale was served. Each pub was identified by a sign with a picture-a Black Horse, White Swan, or Red Lion, for instance. These early "logos" were used because most people could not read. As time went on, the tavern became a permanent institution all over Europe. There were many versions: inns, pubs, cabarets, dance halls, meeting places. Neighbours gathered at these establishments to exchange the latest news and gossip over a mug or a tankard.

In cities, men of similar interests met for a round of drinks and good talk. In London's Mermaid Tavern Shakespeare, dramatist and poet Ben Jonson and other famous literary figures met regularly. Lawyers had their favourite taverns, students theirs. Members of Parliament formed political clubs, each meeting in its favourite tavern for lively discussion of strategy. Whateve: its form, the tavern was a place to enjoy life, to socialize, to exchange ideas, to be stimulated. The beverages intensified the pleasure, loosened the tongue, sparked the wit or, as Socrates once put it, "moistened the soul."

When Europeans immigrated to America, they brought the tavern with them. It was considered essential to a town's welfare to have a place providing drink, lodging, and food. In Massachusetts in the 1650s, any town without a tavern was fined! Often the tavern was built near the church so that parishioners could warm up quickly after Sunday services held in unheated meeting houses. A new town sometimes built its tavern before its church. As towns grew into cities, and roads were built connecting them, taverns followed the roads. In parts of Pennsylvania today, it is possible to find towns named for such early taverns-Blue Bell, Red Lion, Ksing of Prussia.

In some towns, the old tavern is still standing. It was in the taverns that the spirit of revolution was born, fed, and translated into action. These were the rendezvous spots for rebels, where groups like the Sons of Liberty were formed and held their meetings. The Boston Tea Party was planned in Hancock Tavern, while in the Green Dragon, Paul Revere and 30 companions formed a committee to watch the movements of the British Soldiers. In Williamsburg; the Raleigh Tavern was the meeting place of the Virginia patriots, including Patrick Henry and Thomas Jefferson.

In New York's Queen's Head Tavern, a New York Tea Party was planned, and many patriot meetings were held there during the war. After the war, its owner, Samuel Fraunces, renamed it Fraunces Tavern (to shed any reference to the Queen). It was here that General George Washington said good-bye to his fellow officers in 1783. When Washington became President, Fraunces became his chief steward. Today, Fraunces Tavern is a New York City landmark. When Americans pushed westward, taverns sprang up along the routes west. As towns appeared, the tavern was often the first building. Homes and merchants grew up around it.

By the middle 1800s, the "modern" American tavern was becoming a large-scale inn for the travelers and business people of a nation on the move. At the same time, drinking places without lodging were appearing. These kept the name "tavern," while the more elaborate inns adopted the term "hotel." But the hotel kept its bar room; it was often a showplace, with a handsome mahogany bar and a well-dressed bartender who might wear gold and diamonds. Certain hotel bars became famous-the Menger in San Antonio where Teddy Roosevelt recruited Rough Riders, and Planter's Hotel in St. Louis, home of the Planter's Punch.

By the turn of the century, the successors of the early taverns had taken many forms. There were glittering hotels that served the wealthy in cities and resorts. There were fashionable cabarets, such as Maxim's in Paris, where rich and famous men consorted with rich and famous courtesans, and music halls, such as the Folies Bergères. There were private clubs, cafés ranging from elegant to seedy, big-city saloons that provided free lunches with their drinks, and the corner saloons of working-class districts, where many a man drowned his sorrows in drink. The restaurant industry also made its appearance in the nineteenth century, serving wines and other beverages to enhance the diner's pleasure.

Prohibition

Meanwhile; in the United States a growing number of people sought to curb the use of alcoholic beverages. At first this movement went by the name Temperance and its target was "ardent spirits" (distilled spirits). But its proponents soon included beer and wine and expanded their goal from temperance, or moderation, to total prohibition. In a century-long barrage of propaganda and moral fervour, the movement succeeded in convincing many Americans that drink of any kind led inevitably to sin and damnation. If you outlaw "Demon Rum," they believed, sin would disappear and Utopia would naturally emerge. Along with this belief went the notion that those engaged in making or selling alcoholic beverages were on the devil's side of this battle between Good and Evil or, as it was also dubbed, "Dry" and "Wet."

The fervour was fed by the proliferation of saloons opened by competing breweries to push their products, many of them financed by money from abroad. By the late 1800s, there was a swinging-door saloon on every corner of small-town America, as well as in the cities. These establishments often became unsavoury places, because there were far too many of them to survive on sales of beer and whiskey alone, so many became places of prostitution, gambling, and other illegal goings-on.

The Prohibition movement was also an expression of religious and ethnic antagonisms, of fundamentalist middle Americans against the new German and Irish Catholic immigrants. The brewers were German and the bartenders were Irish, and both brought with them cultures that included alcohol intake as a fact of everyday life. The movement also pitted small-town and rural America against what was perceived as big-city licentiousness. During World War I, the Dry side won its battle. The Eighteenth Amendment, passed during the wartime fever of patriotism and self-denial, prohibited the "manufacture, sale, transportation, and importation of intoxicating liquors" in the United States and its territories. Ratified by all but two states, Connecticut and Rhode Island, it went into effect in 1920.

Prohibition had a short and unhappy life-not quite 14 years. There was simply no way to enforce it. While legal establishments were closing their doors, illegal "speakeasies" began opening theirs to those who could whisper the right password. Legal breweries and distilleries closed down, but illegal stills made liquor by the light of the moon in secret hideouts-hence, the nickname moonshine. Illegal spirits also were smuggled into the country from Canada and Mexico and from "Rum Rows" offshore, bootleg supply ships that sold to small, fast boats whose entrepreneurial captains made the run to shore. Some folks just decided to make their own beer, wine, and gin at home.

Prohibition impacted the wine industry as dramatically as it did other alcoholic beverage producers. Many winery owners simply plowed their fields under and planted different crops. A few received special license to make sacramental wines, or permits to make wines strictly for home use, only up to 200 gallons per year. Ironically, rather than decreasing drinking, Prohibition seemed almost to invite it: flouting the law became, to some, the fashionable thing to do. After nine years of Prohibition, New York City had 32,000 speakeasies, about twice as many as the number of pre-Prohibition saloons! To add to the problems of enforcement, organized crime took over the bootleg business in many cities. Gangsters quickly became rich, powerful, and seemingly immune to the law.

The combination of racketeering, gang warfare, and bootlegging became a major national problem. Everyone-even those who first vehemently supported it-agreed that things had gotten out of hand under Prohibition. In 1933, Congress passed the Twenty-first Amendment, repealing the Eighteenth. Before Prohibition shut it down, the beverage manufacturing industry had been the fifth largest in the country. After passage of the Twenty-first Amendment, it made a quick comeback, despite stiff taxes and heavy regulation by federal and state governments.

Today, alcoholic beverages are an accepted part of the American scene, and have been for some time; the sale of liquor is legal in every state and the District of Columbia. The serving of liquor in bars and restaurants is a normal part of the culture, and restaurant patrons expect to be able to buy mixed drinks, beer, and wine with their food. In fact, restaurants that don't serve liquor often have a hard time competing. But the Wet versus Dry controversy never really ended. Control of the issue was given to states, counties, towns, and precincts, resulting in a mishmash of local liquor laws that has made America into a wet-dry checkerboard.

Today's Beverage

Since 1990, there has been a gradual decline in alcohol consumption in all its forms-beer, wine, and spirits. Expert observers relate the drop to a change in lifestyle for many busy Americans, many of whom turned their focus on health, fitness, and job success. These people have stopped smoking, they exercise, they watch their weight and their cholesterol count, and they keep their heads clear during working hours. The "three-Martini lunch" is now a rarity, replaced by bottled waters or fruit juice or sometimes a single glass of wine. These people are moderate drinkers, limiting their consumption to the one or two drinks a day. At the same time, they are very much interested in the quality of whatever drink they choose. When they do imbibe, they tend to choose premium or superpremium liquors and wines. "Drinking less but drinking better" has become the norm.

The simplest kind of beverage enterprise is the bar that serves beverages alone, with no food service except snacks: peanuts, pretzels, cheese and crackers. This type of bar serves beer or wine or mixed drinks or any combination of the three, plus nonalcoholic beverages. It may be a neighbourhood gathering place, a way station for commuters on their homeward treks, a bar at an airport or bus terminal or bowling alley. Business at such bars typically has a predictable flow: a daily pattern of peaks and valleys, a weekly pattern of slow days and heavy days, with the heavy days related to paydays and days off. There may also be seasonal patterns.

In airports and bus terminals, business is geared to daily, weekly, and seasonal travel patterns, and according to the time of day; light beverages are served mornings and afternoons, heartier drinks are ordered as the working day ends. Because only one type of product is sold, and because business is generally predictable, the operation of a beverage-only bar is relatively simple, from production to staffing and purchasing to keeping track of the beverages, money, and profits.

This type of bar also usually has a specific reason for success, perhaps its location, its reputation as a friendly place, or simply a lack of competition; maybe it has just "always been the place where everybody goes." Often such bars thrive by being the same as they always were. Customers become sentimental about them and would not tolerate change. That said, as the mood of the country changes, many neighbourhood bars are adding food to their offerings. Hotel chains such as Marriott, Radisson, and Hyatt have phased out their cocktail-only lounges in favour of food and beverage combinations.

The decision is practical: some states do not allow beverage sales without food sales; other bar owners have decided it's simply more responsible to offer people food if they will be drinking. Master concessionaires, such as Host Marriott, now run more than 1,800 restaurants in 73 airports, and the trend has been to upgrade these facilities to pour more premium beverages, serve better food, partner with brewpubs, and offer entertainment for travelers awaiting their flights. In short, beverage-only bars are definitely a minority today. Although some are highly profitable, most bars find that serving liquor alone is not enough to attract and keep customers. So the majority of bars offer something else-entertainment or food or both.

Entertainment Combinations

Bars offering entertainment range from the neighbourhood bar with pool, pinball, dartboards, or giant TV screens to nightclubs with big-name entertainers and comedy clubs and ballrooms with big bands. In between are cocktail lounges and nightclubs with live entertainment-piano bars, country-and western dancing, jazz or folk duos, or rousing rock-and-roll groups. This concept must include the decision to make room for a stage area, sound system, and dance floor. And having entertainment means hiring someone knowledgeable to book the bands or entertainers that people will want to see and always thinking ahead to the next fad or hottest music trend to attract the fickle public.

A concept that includes regular entertainment of any kind also includes the fixed costs and additional financial risk of hiring and paying the entertainers. In most cases, the entertainment may draw the crowd, but it is the drinks that provide the profits. If there is a cover charge, an admission fee per person paid at the door, at least part of it is likely to go to the entertainers. The fortunes of this type of bar will rise and fall with the popularity of its entertainers, unless the place has something else going for it. Probably the most stable type of bar/entertainment combo is the smaller place with an attractive ambience, good drinks, and local entertainment to draw a loyal, local crowd. Its success potential is much the same as the baronly enterprise. Larger operations featuring out-of-town entertainers have a higher but riskier profit potential. It is likely to be either feast or famine.

The bar gears up for each crowd with temporary extra help, a large investment in liquor inventory, and possibly extra security personnel. Weather, holidays, location, and weeknight versus weekend crowds all heavily impact this type of business. Casinos are another enduring combination of entertainment and beverage service. Today's casinos may be run by a huge corporation or a Native American tribal council, and may include everything from big-name stage productions and professional boxing matches to restaurants and nongambling arcades that attract families instead of adults only.

Sports bars offer a different type of entertainment. They center on the viewing of popular sporting events, such as Monday Night Football, or special events like hockey finals or the World Series. Equipped with large television screens, the sports bar often sets a fixed price or cover charge to guarantee a good profit because customer turnover is so small. Large sports bars often serve a menu of full-course meals, and many take reservations in advance of popular events like a professional boxing match or a Triple Crown horse race. The cigar bar is another trendy addition to the beverage scene-and a profitable one, too. Men and an increasing number of women are enjoying high-priced cigars, and restaurateurs have seized the opportunity to recommend premium spirits, wines, beers, and after-dinner drinks to accompany them.

The cigar boom is not legal in all venues, since smoking is prohibited in many public places by local ordinance. But the places that install heavy duty fans and humidors and offer extensive cigar selections-plus single-malt Scotches, small-batch Bourbons, Cognacs and Ports-are filling an interesting, upscale niche. Sometimes, they are private clubs that charge membership fees.

Food and Beverage

The most common form of beverage operation is one that is linked with some kind of food service. One type is the restaurant/bar, where drinks and wine are part of the meal service, served by the same wait staff that serves the meal. The bar is often used as the waiting area for the restaurant during busy times. Drinks may be poured at a service bar out of public view or at a pickup station in a bar that serves customers while they are waiting for a table. The major portion of the sales comes from the food service. However, the beverage sales often turn the profit for the enterprise. The only added costs are for the wine and liquor, and the bartender and a minimum investment in equipment; the other necessities-service personnel and the facility itself-are built in to the restaurant operation.

Another type of food-beverage combination is the bar that offers light food in addition to drinks. In this case, the beverages and the bar atmosphere dominate, and the major sales volume comes from the bar. But the food is a nice sidelight that attracts customers and prolongs their stay. Typical menu items are appetizers: nachos, chips or crudités and dips, spiced chicken wings, stuffed potato skins. A special variation of the food-beverage combination is the wine bar, which first appeared during the 1970s as Americans discovered and learned to appreciate wines.

Here the customer can choose from a selection of wines by the glass or by the bottle, beginning with inexpensive house wines and going up in quality and price as far as the entrepreneur cares to go. Some wine bars offer inexpensive one-ounce "tastes" to allow guests to sample a number of wines. A full menu can be served, or fruit and cheese platters and upscale hors d'oeuvres.

There are inherent problems in running wine bars. The first is, of course, that serving only wine tends to limit the clientele to wine lovers. In some urban areas, there are enough wine enthusiasts to support a profitable enterprise; they respond to quality and expertise, and they attend and appreciate special tastings and classes and wine-centered celebrations. This enthusiasm, however, raises a second difficulty: purchasing appropriate wines requires an expertise few people have, and may require a financial investment few are willing to make.

As a result, many wine bars serve liquor and beer as well. This broadens their appeal and allows them to realize the necessary profit margin. In effect, they are simply bars that specialize in wine sales and wine knowledge. Other wine bars may broaden their offerings by serving meals, in effect becoming restaurants with an emphasis on wines. Some also sell wines at retail, offering customers discounts for volume purchases. This combination of on-premise service and take-home sales is not an option everywhere. Beverage laws in many areas do not allow it.

Beer aficionados also have their own version of the wine bar. At a brewpub, beer is brewed and served right on the premises-fresh, natural beers and ales, strong in flavour and aroma, with special seasonal offerings. Developed by small individual entrepreneurs and hobbyists, the beverage sets the theme of the restaurant. At least one shiny brew kettle is likely to be a major part of the décor, and the menu typically contains hearty, casual cuisine chosen to complement the beer. As popular as they are in many areas of the United States, brewpubs are not legal everywhere, as some states still do not allow manufacture and sale of alcoholic beverages on the same premises.

A popular type of food-beverage combination links a bar and a restaurant on an equal, semi-independent basis, with a common roof, theme, management team, and services that complement each other. The bar and restaurant areas are housed in separate portions of the building, and they may be open at slightly different hours to serve both the drop-in bar customer and the meal time patron. The food/drink sales ratio is likely to reflect an equal status of food and drink, with bar and restaurant each doing better than it would without the other. In many cases, neither side could make a go of it alone, but together the customer attraction and income are doubled, while the overhead costs are split between them.

Beverage Operations in Hotel

In hotels, the beverage operation differs in many ways from the bar or the bar-restaurant combination. There may be three or four bars under one roof, each with a different purpose and a different ambience-say a lobby bar, a cocktail lounge, a restaurant bar, a nightclub with dancing. In addition, there is room service, with a food menu that includes mixed drinks, beer, wine, and Champagne. Above all, there is banquet service, catering to conference, convention, and reception needs. Typically, the client makes beverage choices in advance of the event, which are served from portable bars by extra personnel hired for the occasion.

Many hotels have installed them for the convenience of guests, an in-room refreshment center without the room service waiter. The minibar is really nothing more than a small refrigerator with an icemaker-although many have classy mock mahogany exterior cabinets-stocked with a small inventory of snacks and drinks. There are three keys to profitable minibar use, according to Lodging magazine. The unit must be installed so that it is easy to use and its contents are clearly visible, and a reliable system must be in place for prompt restocking of cabinets and correct billing of guests. Perhaps most important, customers must be enticed to somehow overlook the high prices of minibar goods. Its in-room convenience means the hotel can charge double, or even triple, what the same goods would cost elsewhere on the same property.

Food and Beverage Directors of large hotels say the minibar has become a necessary amenity, even though it could be argued that it does siphon some business away from the hotel's other food and beverage venues, especially room service. But overall, the minibar is not a major money maker for most hotels; and, interestingly, the most common minibar purchase is not alcohol of any kind, but bottled water. Perhaps the most daunting challenge of hotel beverage service is its diversity, coupled with the up-and-down nature of demand. Since a hotel's primary clientele is its overnight guests, demand for beverages rises and falls according to the occupancy rate. This, too, is unpredictable: a hotel can be completely full for a convention and yet have very little bar trade, depending on the kind of convention it is hosting.

On the other hand, a very low occupancy rate may net a lot of bar business. Again, it just depends on who the hotel guests are. Resort and luxury hotels often have several bars and restaurants, with a variety of entertainment, food, and drink, to keep the hotel guests spending money on the premises, as well as to attract an outside clientele. On the other hand, a small commercial hotel in a big city may need only one bar with several stations to serve its lobby customers, its cocktail lounge, coffee shop, dining room, and room service.

Beverage Service in Airline

Another type of beverage service that must adapt to special conditions is that on airline flights. The restrictions of space, time, weight, and equipment are formidable. Of necessity, their drink menus are limited. Liquors, beers, wines, and a few types of cocktail mixes are handed out in small individual bottles or cans. The cups are nesting, plastic disposables, except in first and business class cabins. Flight attendants push a beverage cart down the aisle and, working from both ends, can garnish glasses and fill them with ice, pour beverages or hand out the individual-sized drink components, and collect the money.

The process is a marvel of organization. Tight control systems follow the little bottles everywhere, since they are extraordinarily tempting to both airline employees and customers. For higher-paying passengers, drinks are free and service typically includes real glassware, a choice or wines, Champagne for breakfast, and sometimes specialty drinks.

Similarities and Differences

Grouping types of beverage service into these rather arbitrary categories does not really adequately describe the character of individual enterprises. Many establishments do not fit handily into a specific category, and those within categories can be as different as day and night. Yet all categories have certain similarities. They all sell alcoholic beverages. They have similar staff structures, patterns of purchasing and inventory, and ways of controlling the merchandise. They all must meet certain government requirements and operate within certain government regulations.

Even the prices charged for the same drinks are not wildly different from one type of place to another. Still, no two bar and beverage operations are alike, unless they are part of a chain. The successful business is one that meets the needs and desires of a certain clientele and strives to be deliberately different from others serving a similar clientele in order to stand out in the competition for customers. Other major reasons for the wide variety of bar operations are simple: the special circumstances of each operation and the personalities, desires, and budgets of their owners. But to be successful, the entrepreneur must put clientele above all else in shaping his or her enterprise.

Numerous programs exist for the food and beverage service industry that help restaurants and their customers minimize environmental impacts. The Green Restaurant Association's approach as an outline to present the environmental practices a restaurant should consider.

Green Restaurant Association (GRA)

Involvement in environmental programs is a good way for a restaurant to learn more about environmental best practice opportunities. An environmental resource for the food service industry is the Green Restaurant Association (GRA), which helps member restaurants reduce environmental impacts with twelve "eco-steps". The GRA's primary operational components are research, environmental consulting, education, public relations and marketing, and community organizing and consumer activism.

To become a member of the GRA, a restaurant must sign a statement of its commitment to making environmental improvements based on the 12 "eco-steps" listed below. After a member restaurant has made positive environmental changes, the GRA will include the restaurant in the "Green Restaurant Guide," which serves as a directory for environmentally conscious restaurants. The GRA also offers fee-based consulting services to member restaurants and initiates public relations and marketing initiatives, such as having restaurants featured on CNN, to increase restaurant exposure and consumer awareness.

GRA's 12 eco-steps are:

— *Elimination of Polystyrene Foam (commonly known as styrofoam)*: Replace all polystyrene foam products with environmentally friendly alternatives: paper, bamboo or sugarcane paper, recyclable plastic, biodegradable plastic, and so on. This is the minimum environmental standard for becoming a member of the GRA.

— *Comprehensive Recycling*: Initiate or improve recycling programs for glass, plastic, bimetal, cardboard, and mixed paper.

— *Waste Reduction and Reuse*: Increase bulk purchasing and reduce excessive packaging for food, condiments, and so on. Replace disposable products with reusable alternatives: eating ware, aprons, tea strainers, cups, and so on.

— *Biodegradable Plastic*: Transition to corn-based "plastic" products that are biodegradable and petroleum-free: cups, utensils, garbage bags, and straws.

— *Recycled Products*: Transition to recycled products with the highest postconsumer content available and non-tree-fiber paper products: napkins, paper towels, toilet paper, office paper, take-out containers, coffee jackets, plates, and bowls.

— *Non-Chlorine -Bleached Paper Products*: Transition to non-bleached or non-chlorinebleached paper products: cups, wax paper, plates, take-out containers, bags, pastry bags and grabbers, napkins, paper towels, coffee filters, and office paper.

— *Nontoxic Cleaners, Landscaping and Pest Management*: Replace hazardous chemical products with biodegradable and nontoxic alternatives; dish detergent, germicides, disinfectants, toilet bowl cleaners, drain cleaner, floor wash, floor polish, glass cleaners, degreasers, and laundry detergent. For landscaping, switch to nontoxic, nonsynthetic, and organic fertilizers, pesticides, or herbicides. For pest control, use nontoxic products or services.

— *Energy Efficiency* : Improve the energy efficiency of lighting, refrigeration, air conditioning, gas appliances, and so on. Obtain assistance with using government and private rebate programs. Make connections with energy consultants to learn about more extensive programs.

— *Water Efficiency* : Improve the water efficiency of toilets, faucets, laundry, sprinkler systems, and so on.

— *Composting* : Divert food waste from landfills and create nutrient-rich soil for gardening and landscaping.

— *"Green" Electricity* : Change to an energy pr ovider that uses solar, wind, small-scale hydroelectric, geothermal, or methane-based power that is renewable and less polluting to ecosystems.

— *Employee Education Program* : Train all employees, managers, and owners. Topics covered should include

  — An environmental profile of the restaurant industry

  — A history of environmental issues relevant to food service: landfills, water pollution, air pollution, clear-cutting, and global warming

  — Data describing the restaurant impacts (positive and negative) on the environment

Polystyrene (Styrofoam) Foam

Polystyrene foam (Styrofoam) is widely used in restaurants in insulated cups for hot beverages and take-out containers. Production of polystyrene involves use of known (benzene) and suspected (styrene and 1,3-butadiene) human carcinogenic substances (styrene and 1,3-butadiene). Styrene is also known to be toxic to the reproductive system. Polystyrene can be recycled; however, its recycling rates are low. Styrofoam is light in weight, but bulky in size, so hauling Styrofoam to the nearest available recycling facility is often not economically feasible.The value of a load of Styrofoam may not even cover the cost of shipping. To reduce environmental impacts, restaurants should consider switching from Styrofoam to non-bleached paper wraps, cardboard containers, or other sustainable food packaging.

According to a report published by the Environmental Defense Waste Reduction Task Force, McDonalds has completed the switch from polystyrene foam "clamshells" to paper-based wraps for packaging its sandwich items. The wraps provide a 70 to 90 percent reduction in packaging volume, resulting in significantly less space being consumed in landfills. Compared to the polystyrene foam boxes they replaced, the new sandwich wraps also offer a substantial savings in energy used and substantial reductions in pollutant releases measured over the full life-cycle of the packaging. McDonalds is also testing EarthShellpackaging products in 300 of its restaurants throughout the U.S. These sustainable containers are made from potato starch, natural limestone, 100 percent postconsumer recycled fiber, biodegradable polymer and wax coatings, and water.

Recycling Program

The "Recycling Guidebook for the Hospitality and Restaurant Industry" provides a general overview of developing and implementing a comprehensive recycling program.The first step mentioned in this resource is to conduct a waste audit in order to evaluate the waste stream, enabling a restaurant to better target the commodities that should be included in a recycling program. Materials that are commonly recycled in restaurants include

— Paper, including cardboard, computer paper, register tape, and telephone books
— Metals, including aluminum, tin, and steel cans
— Green, brown, and clear glass

— #1 Polyethylene (PET) and #2 high-density polyethylene (HDPE) plastics

— Printer cartridges

Convenience is the key to a successful recycling program. An EPA (Environmetal Protection Agency) Waste Wise tip sheet notes that a convenient collection system will encourage both customers and employees to carefully sort recyclables by material type and to eliminate contaminants. Collecting uncontaminated recyclables (commodities that are properly sorted and free of excess food and beverage waste) will save time otherwise spent in sorting out contaminants. Further, uncontaminated recyclables have higher value if they are sold. Following are other restaurant recycling tips listed in the "Recycling Guidebook for the Hospitality and Restaurant Industry."

— Recycling and trash bins should look different from each other and be clearly marked. Both types of bins should be conveniently located in the kitchen and bar areas so that employees will use them.

— Self-serve establishments should post signs to inform customers about the recycling program and provide specific instructions. Either strategically places bins for collection of recyclable commodities, or have customers leave such commodities on a designated counter for collection by staff.

— If lack of space is a problem, specially designed equipment such as can, glass, and plastic crushers are available to reduce the volume of recyclable materials.

— Recyclable collection bins in public areas should be well-marked. Choose bins with specialized openings, such as a hole for cans or a slot for newspapers, for these areas.

— Set up a logbook or a receipt system to record the volume of recyclables leaving the premises in order to facilitate tracking and compensation.

Waste Reduction and Material Reuse

Waste reduction and material reuse should be the first step in minimizing the waste that a restaurant produces because it is more efficient to reduce waste at the source or reuse material than it is to recycle. Conducting a waste audit is an effective way to evaluate a restaurant's waste streams. Doing so enables staff to identify wastes that are nonessential to operations, such as excess packaging material. This process can also identify disposable products that can be replaced with durable, reusable ones. For example, disposable plastic tableware and silverware should be replaced with washable or compostable utensils.

EPA's Waste Wise program is a free, 3-year, goal-oriented program that assists businesses in assessing and reducing their waste streams.EPA designed Waste Wise to be a flexible program, in which the participant determines how much time and money to invest. Key aspects of successful Waste Wise programs include

— Waste assessments

— Employee education

— Measurement and reporting

— Program maintenance

The above aspects are fundamental in assessing, developing, implementing, and maintaining a viable waste reduction and reuse program. Waste assessments provide a full understanding of waste streams and provide a basis for targeting specific waste reduction goals. However, these goals will only be reached with the full support of the restaurant — from managers and servers to cooks and food preparation staff. Therefore, employee education is critical to the success of the program. It is also important that program efforts be measured regularly. Tracking and documenting the costs, savings, and effects of a waste reduction and reuse program are the only ways to determine the environmental and economic impacts of implementing the program.

Program data can also provide insight into the effectiveness of employee education programs. Lastly, program maintenance is imperative to realize long-term benefits from a waste reduction and reuse program. A successful program is not a result of one-time changes but a cumulative result of permanent procedural and behavioral changes. Several other guidance documents provide specific waste reduction and reuse ideas for the restaurant industry. Following are some examples of common waste reduction techniques:

— Purchase and serve beverages in bulk rather than in bottles, cans, or individual packets.

— Buy bar mixes, juices, and coffee in bulk.

— Use health department-approved, refillable condiment containers.

— Purchase cleaning supplies in concentrated form.

— Use cloth towels and reusable table linens and tableware.

— Use vendors that take back packaging material and pallets.

Another way that restaurants can minimize their waste and overall environmental impact is purchasing products through local food supply vendors and grocers. Local sourcing is of particular benefit to the tourism sector, where visitors appreciate locally distinctive food, drink, and other products. For example, several Aspen Skiing Company-owned restaurants purchase natural, chemical-free beef and hamburger from local ranchers to support the ranching community and thus preserve open space, as well as to provide guests with healthy food. The pilot program began in winter 1998 and expanded significantly in 2000.

The beef costs twice as much per pound, however Aspen Skiing Company absorbs the additional cost. By purchasing local produce and supplies, restaurants minimize the affects of excessive transportation and shipping. Such efforts should be accounted for when evaluating the environmental impact and overall performance of a restaurant.

Biodegradable Plastic

As long as customers wish to take their food "to go," restaurants will need to stock disposable goods such as food containers and silverware. According to Biocorp, a manufacturer of biodegradable plastic tableware and silverware, nearly 113 billion disposable cups, 39 billion disposable eating utensils, and 29 billion disposable plates are used in the U.S. every year, and half these items are made of plastic. Because restaurants cannot control what happens to these items once they are in customers' hands, the most effective way to minimize their environmental impacts is to purchase biodegradable products.

Biodegradable products are typically made of corn, starch, or paper with an easily biodegradable coating. The EarthShell is one example of a biodegradable food service product. Biocorp also manufactures biodegradable composting bags, silverware, plates, and beverage containers."Scientists Perfecting Planet-Friendly Plastics," an article published by the Environmental News Network, explains that the starch used to create these biodegradable plastics - typically wheat gluten -costs about 15 cents per pound, whereas the least costly commercial plastics cost about $1 per pound. Thus, when this starch-based plastic becomes widely available to manufacturers, it could be the cheapest plastic available.

Currently, however, most manufactured environment-friendly plastics cost about $2.50 per pound, although recent projects have brought their costs down to approximately $1.50 per pound. Because biodegradable plastics are more expensive than regular plastic, the biodegradable plastic industry has a challenge in breaking into a very competitive market.

Recycled Materials

Restaurants use several types of products that can be replaced with items made from recycled materials. Most of the products with recycled content that can easily be purchased are paper products such as napkins, toilet paper, tissues, paper towels, paper tablecloths, take-out containers, register tape, and office paper. Since 1998, EPA has required federal facilities to purchase products with recycledcontent material. The resulting increased demand for recycled content paper has driven down its cost to the point that recycled-content products are the same price or less expensive than virgin material products.

EPA's "Comprehensive Procurement Guidelines" set recommended recovered-material content ranges for paper products. Restaurants should purchase paper products that fall within EPA's recommended ranges. Paper product suppliers have information about the recycled content of the products they carry.

Bleached Paper Products

Restaurants should avoid purchasing Non-Chloring-Bleached paper products. Seventh Generation, a vendor of environmentally friendly, nontoxic consumer products, explains that chlorine is used by the paper industry for two purposes: to dissolve lignin, a natural material that holds a tree's cellulous fibers together, and to whiten the final paper product. However, chlorine reacts with the virgin natural substances of trees and recycled paper material to form both dioxins and organochlorines. As these substances are discharged and accumulate in the environment, they can have profound long- and short-term health effects on exposed humans and wildlife.

One concern regarding dioxins is that they are known carcinogens, or cancer-causing substances. Short-term reactions to overexposure to chlorine may also occur, including airway inflammation and bronchial hyperresponsiveness. Organochlorines are of concern because they are suspected to be endocrine modifiers, which act as hormones in the body and can disrupt the human immune system.

Nontoxic Cleaners, Landscaping and Pest Management

Restaurant personnel use a wide variety of chemicals to clean kitchen, dishwashing, and restroom facilities. Some cleaning products used include toilet and tile cleaner, glass cleaner, carpet cleaner, spot remover, disinfectant, and oven cleaner. Many of these products contain chemicals that are harmful to human health and the environment, which is a concern for both restaurant staff and customers. Vendors offer environmentally preferable cleaning supplies with equal or better cleaning performance at equal or less cost. Many restaurants have outdoor areas with landscaping. There are several relatively simple steps that a restaurant manager can take to minimize the environmental impact of maintaining these areas, including water conservation and selection of climate-appropriate indigenous plants. Some "green" practices for landscape management include

— Watering vegetation using "deficit" irrigation, or frequent light watering
— Selecting plants based on watering needs (Typically, indigenous plants are most efficient in using water.)
— Watering grounds during the coolest hours of the day (typically at night) Finally, pest management can also be an issue for restaurants.

Efficiency of Energy

Most restaurant operations, such as cooking and cleaning,

are energy-intensive. EPA's Energy Star Program estimates that reducing energy consumption by 20 percent can increase a restaurant's profit by one-third, a good economic incentive to make energy efficiency investments. Making this sort of reduction is feasible when restaurants implement strategic energy conservation measures. The Energy Star Program analysis of restaurant energy use shows that over 60 percent of a restaurant's energy is used for cooking, heating water, and heating the establishment. Following are the Tips for reducing energy costs for restaurants.

Turn on and off

— Preheat equipment in accordance with manufacturer specifications; post preheating times near equipment.
— Turn off appliances when not in use.
— Install occupancy sensors in walk-in refrigerators and freezers.

Cooking efficiency - Lower it / fill it up

— Cook using equipment at full capacity when possible.
— Cook at lowest temperatures first.
— Turn off equipment during downtime.
— Use lids to minimize heat loss.

Watch thermostats - Stay cool and save money

— Set thermostats to manufacturer-recommended temperatures.
— Apply new-generation "clear" coatings to reduce solar gain from large, south- and west- facing windows.
— Use the "unoccupied" and "night setback" thermostat options.
— Ensure tamper-proof temperature settings by using locking covers on thermostats.

Keep it clean for energy efficiency

— Clean condenser and evaporator coils on air conditioning and refrigeration equipment.

— Change all filters regularly.

— Preventively maintain equipment.

Dishwashing equipment

— Heat water only to the temperature required for specific tasks.

— Install equipment of proper size.

— Fully load the machine for each cycle.

As restaurants undergo renovation or technology upgrades, managers should evaluate the energy efficiency of new equipment. Green Seal is an independent, nonprofit organization dedicated to protecting the environment by promoting the manufacture and sale of environmentally responsible consumer products. Green Seal sets environmental standards and awards a "Green Seal of Approval" to products that cause less harm to the environment than other, similar products.Commercial consumers can use Green Seal's web site to search for resource efficient products. A similar feature is offered on EPA's EnergyStarProgram web site.

Efficiency of Water

Restaurants use water in almost every aspect of their operations, including food preparation and cooking, cleaning, in restroom facilities, and as a beverage for customers. For each of these aspects, there are technologies or best management practices (BMP) that conserve water. In a collaborative effort, Agricultural and Biological Engineering, EPA, and Purdue University created an environmental enrichment toolkit for the lodging industry. Following are water conservation tactics for restaurants:

— Wash food products in buckets, bowls, or other containers.
— Only run dishwashers with full loads.
— Regularly inspect dishwasher pumps for water leaks.
— Defrost or thaw frozen foods in the refrigerator instead of water.
— Install low-flow taps in kitchens and restrooms.
— Use low-flow toilets in restrooms.
— Immediately fix any leaking or dripping faucet.
— Install infrared-activated faucets and toilets in restrooms.
— Purchase and use water-saving kitchen equipment.
— Track water consumption by regularly monitoring utility bills.
— Establish an effective employee training program on water conservation.

Composting Programs

EPA estimates that food wastes comprise 6.7 percent by weight of the total U.S. municipal solid waste stream.One option for diverting food wastes from landfills is composting. Businesses with well-established composting programs divert 50 to 100 percent of their food scraps and reduce their overall solid waste by 33 to 85 percent.Composting can be done both on and off site; however, for restaurants where space is limited, the most feasible option is to collect food scraps for an off-site composting program. This option also transfers the responsibility of monitoring the chemical balance of the compost from the restaurant owner to the local composting operation. However, technologies are available that enable restaurants to manage their compost on site.

Keystone Ranch, a horse ranch resort in Keystone, Colorado, implemented a composting program that combines horse manure with vegetable and meat scraps from the Ranch Restaurant.The ranch rents a 30-cubic yard container from a local hauler that holds almost 20 tons of material. Biodegradable cornstarch-based bags are placed at four stations to collect food waste during food preparation and uneaten food from customers. Keystone Ranch estimates that between 10 and 15 percent of the contents of the full 30-cubic yard container is food waste. The rest of the compost mixture is manure collected from the stalls of 15 to 80 horses. The hauler transports the container to the Twin Landfill Corporation in a nearby town, where the compost is further processed, screened, and then sold to landscapers and farmers.

As a result of the composting program, which complements an established glass, aluminum, tin, cardboard, and paper recycling program, the Ranch Restaurant is "rapidly approaching a zero-waste status." Because the tipping fee for compostable material is 30 percent less than that for solid waste, the program sustains itself financially and has diverted 450 tons of organic waste from the local landfill.

Green Electricity

In addition to efforts to minimize their energy use, restaurants can minimize the environmental impacts of the energy they do use by purchasing renewable electricity. By using a renewable source of energy, a restaurant decreases its dependency on nonrenewable fossil fuels and eliminates greenhouse gas emissions associated with its electricity use. Renewable energy sources include wind, solar, geothermal, and biomass.

The availability of green electricity varies from region to region, so restaurants interested in purchasing energy from a renewable source should contact their city or state government agencies or local electric utility. Non-profit green energy certification programs, such as Green-e and Cleaner and Greener, may also be able to help direct restaurant owners to local green energy providers.

Education of Employee

The development of an employee education program is critical to maintaining an environmentally responsible restaurant. An environmental education program should teach both the "how" and "why" of each aspect of the program. An understanding of why an employee should ensure that certain environmental procedures are followed adds meaning to the task. Because of high turnover in many restaurants, environmental education should be a regular, multimedia program, and should include both written and verbal instruction in all languages of the employees. Environmental awareness expectations should be outlined during new-hire training and reinforced throughout the period of employment.

Restaurant managers should consider acknowledging employees who demonstrate superior dedication to minimizing environmental impacts. For example, employees may be offered a percentage of the cost savings resulting from environmental projects implemented at their suggestion. Such recognition would not only reward the employees, but would send a message to customers that the restaurant is actively pursuing "greener" practices. Once the expectation of excellent environmental performance is established, maintaining the program will require less effort. Finally, restaurant managers should encourage employees to provide feedback regarding areas for improvement. Many employees will have ideas of how to improve the environment they work in, and with a forum to voice ideas, innovations can be made.

Bibliography

Briggs R., Food *Purchasing and Preparation*, Continuum Books, 2000

Buttle F., *Hotel and Foodservice Marketing*, Continuum, 1996

Conran T., *Terence Conran on Restaurants*, Eddington Hook, 2000

Cousins J, Foskett D & Shortt D., *Food and Beverage Management*, Longman, 1995

Dashwood, C., *The Substantive Law of the EEC*, Sweet and Maxwell, 1997

Dix C & Baird C., *Front Office Operations*, Longman, 1998

Harris P., *Profit Planning*, Butterworth Heinemann,1999

Johns N., *Productivity Management in Hospitality and Tourism*, Continuum Books, 1995

Jones P & Lockwood A., *Management of Hotel Operations*, Continuum, 1999

Jones P & Merricks P., *Management of the Food Service Operation*, Continuum, 1994

Julyan B., *Sales and Service for the Wine Professional*, Continuum Books,1999

Kirk D., *Environmental Management for Hotels*, Butterworth and Heinemann, 1996

Knight J & Kotschevar L, *Quantity Food Production, Planning and Management*, 2000

Knowles T., *Hospitality Management an Introduction*, Longman, 1998

Kotas, R. & Chandana, J. Profitable Food & Beverage Management, 1995

Lashley & Morrison, *In Search of Hospitality,* Butterworth and Heinemann 2000

Lashley C, Morrison A, *Franchising Hospitality Services·* Butterworth and Heinemann, 2000

Lockwood A, DavisB & Stone S, *Food and Beverage Management* BH 1998

Macmillan S. Tuxworth Twist W., *Environmental Management for Hotels,* Butterworth and Heinemann, 1996

Mullins L., *Managing People in the Hospitality Industry,* Longman,1998

Rochat M., *Food and Beverage Marketing,* Eddington Hook, 1999

Waller K., *Improving Food and Beverage Performance,* Butterworth and Heinemann, 1996

Wood R, *Strategic Questions in Food and Beverage Management,* BH, 2000

Index